PSYCHOLOGY RESEARCH PROGRESS

PSYCHOANALYSIS AND COMPLEXITY

PSYCHOLOGY RESEARCH PROGRESS

Additional books in this series can be found on Nova's website
under the Series tab.

Additional e-books in this series can be found on Nova's website
under the e-book tab.

PSYCHOLOGY RESEARCH PROGRESS

PSYCHOANALYSIS AND COMPLEXITY

GABRIELE LENTI

nova publishers
New York

For permission to use material from this book please contact us:
Telephone 631-231-7269; Fax 631-231-8175
Web Site: http://www.novapublishers.com

Library of Congress Cataloging-in-Publication Data

Psychoanalysis and complexity / editor, Gabriele Lenti.
pages cm. -- (Psychology research progress)
Includes bibliographical references and index.
ISBN: 978-1-62948-318-4 (hardcover)
1. Psychoanalysis. 2. Complexity (Philosophy) 3. System theory--Philosophy. 4. Psychoanalysis and philosophy. I. Lenti, Gabriele.
BF175.4.P45P53 2014
150.19'5--dc23

2014017743

Published by Nova Science Publishers, Inc. † New York

To my wife, Sabrina

Anyone who engages in the scientific development of hypothesis
begins to take in serious consideration their theories
only when they can be inserted in knowledge
from more than one point of view

Sigmund Freud

CONTENTS

Foreword	*Beppe Berruti*	**ix**
Preface		**xi**
Chapter 1	At the Beginning	**1**
Chapter 2	Myths Dispelled	**7**
Chapter 3	Evolution and Nonlinearity	**15**
Chapter 4	Relation versus Reification	**23**
Chapter 5	Causality	**35**
Chapter 6	Loops	**39**
Chapter 7	Data Complexes	**43**
Chapter 8	Cause and Words	**45**
Chapter 9	In the Beginning Was the Uncanny	**49**
Chapter 10	The Emergence of Order	**53**
Chapter 11	The Arrow of Time	**55**
Chapter 12	Open Systems	**57**
Chapter 13	Sensitive Dependency on Initial Conditions	**59**
Chapter 14	Complexification	**61**
Chapter 15	Possibilities	**63**
Chapter 16	Turbulences	**65**
Chapter 17	Strange Attractors	**67**
Chapter 18	Boundless Objects	**71**
Chapter 19	Complicated or Complex?	**75**
Chapter 20	New Organizational Equilibria	**81**

Chapter 21	The Subject-Object	**83**
Chapter 22	Feedback	**85**
Chapter 23	Autopoiesis	**87**
Chapter 24	Relationships and Analytical Field	**91**
Chapter 25	The Concept of Field and Complex Logic	**95**
Chapter 26	Transformations in the Analyst's Consulting Room	**99**
Chapter 27	Learning	**103**
Chapter 28	The First Steps of Thought	**105**
Chapter 29	Projections	**109**
Chapter 30	Internal Projections	**113**
Chapter 31	Holoprojections	**117**
Chapter 32	Bion's Signs As Operators of a Complex Language	**119**
Chapter 33	More on Hologrammic, Recursive and Dialogic Organization of the Field	**123**
Chapter 34	The Analytic Field and Nonlinearity	**127**
Chapter 35	Promoting the Nonlinearity of Transformations	**139**
Chapter 36	Interactions	**141**
Chapter 37	Naming	**143**
Chapter 38	Epistemology and Style	**145**
Glossary		**157**
Bibliography		**159**
About the Author		**169**
Index		**171**

FOREWORD

Beppe Berruti
Analyst/Psychologist, Psychiatrist
Mental Health Department of The Local Health Board, Savona, Italy

Just a few words to introduce this work by Gabriele Lenti; a few words, since complexity, which Lenti so deeply explores, requires precision and rejects overlapping and multiplication of unnecessary words; a few words from the clinician's point of view, which refers and must refer to models and theories in their therapeutic activity.

The clinician has always been torn between an apparently exhaustive and linear theory and their uncertain practice, which alters and belies the certainties of reference models.

In this sense, every day each of us alters the applied theories and classifications still mostly continuing to use them. Moreover, psychoanalysis has always been a step back or a step ahead of science in the epistemological field, or however it has been looking for an organic relationship with it.

Therefore it is clearly understandable that we therapists are particularly aware of this Procrustean bed and consider complexity a field which, though belonging to the "great" sciences of Nobel prizes, nevertheless seems aware of our problems, our style and the research method we apply daily.

Theorization about complexity has actually broadened our view over previously unthinkable fields, while psychology, which has always been engaged on this front may feel attracted to and benefit from this confrontation, as the reader will understand by reading this book written by this colleague who is not afraid of testing his double nature of therapist and philosophic researcher.

Still all that glitters is not gold and idealization is always dangerous; in modern times of an explosion of information and literary production this is particularly true in a scientific field which has dropped a univocal point of view and a unique direction, hence multiplicity not only of theories but also of rhetoric and difficulty in confrontation may confuse the researcher (and in this psychoanalysis is doubtless still a step ahead).

For this reason and because complexity informs our activity as clinicians (and when we give it up we do no good to ourselves and to our patients) all contributions, which with the right amount of competence, determination and humbleness deal with the reasons of our everyday work, are important; otherwise it often consists merely of a thoughtless practice, even if we act with the best of intentions.

Gabriele Lenti has tried all that and proposes a contribution of his, which will surely provide us theoretical references and more in our activity as therapists.

PREFACE

Psychoanalysis is one of those knowledge fields most frequently explored through conceptually hybrid tools which are half new, as its complexity requires, and half obsolete since its cultural heritage tends to reify its descriptive language which is necessarily metaphorical.

Furthermore it seems that psychotherapy and psychoanalysis in particular, have produced with the proceeding of knowledge also a methodological system aimed at *objectifying* the reality they are studying.

Through the expression "*objectifying* the reality" I mean something different from "reifying the language"; I mean that process of arbitrary selection of what must be considered the *object* of study.

The reality we study is subject to a previous *objectifying purge,* to the extraction of those characteristics which can more easily processed by the particular theoretical models we choose to apply.

Science is no more a purifying queen, due to the nature of its object of study; it is rather a humble servant leading us to a constantly temporary, relative knowledge which cannot afford the luxury of thinking it is self-referential, at least from Gödel forwards.

The knowledge of the object is indeed a function of the observer's cognitive tools in analytic phenomena; consider, for example, the coupled unconscious fantasies; subject and object are so co-substantial to require a paradigmatic epistemological jump.

It is not as important anymore to recognize objectifications of personality as to grasp *relational constellations d*escribing the dynamics of the emotional field which involve all the actors, real or fantasized, on the analytical scene.

These constellations are caught by words, emotions and complex narrations which interweave the discourse, and also by the images and by the characters in order to describe and at the same time lead the object to openness to new meanings.

It is a temporary though important achievement as from here relationship can evolve and the meaning can recursively retroact on the constellations, changing them, transforming suffering into an experience of growth.

This way of conceiving and observing natural phenomena introduces another element of novelty: it is not possible anymore to think according to a linear logic, since recursion originates transformations which proceed through disproportions between causes and effects, sudden changes, chaos, unpredictable bifurcations. This is a nonlinear logic, a complex generative dynamics, where projective, introjective and restitutive interactions are causes and producers of the field by which they are produced.

The organization of the field is therefore different from the mechanical order based on the Aristotelian logic. The field requires principles which do not absolutely (ontologically) oppose each other but rather describe an antinomic simultaneity between order and chaos, expression of a deep natural symmetry, of a specific invariant of the complex.

Moreover we will see that the emotional field has a Hologrammatic organization, as the whole (emotional field) and the part (elements of the internal world) are similarly organized, include each other and reciprocally retroact in a game of endless projections.

This new epistemological approach results also in a particular linguistic style used in the description of the events during the session, a style containing a large amount of literary devices and yet sober and essential as a running commentary.

The evolutive potentiality of every analytic intervention is indeed as evanescent as the present experience; the language of the individual myth is rewritable in "here and now", sometimes collectively, however always hovering between sharing and regressing to the pleasure of the primary process.

"Through the affective-cognitive experience of analysis we cannot change the basic conditions but take up and promote a symbolic activity which permits a sort of weaving *inside* the traumatic wound, making it bearable and creative" (my emphasis) (Algini 2002).

In this passage I have purposely substituted the original *around* with *inside* in order to underline the poietic and systemic nature...

My considerations are intended to be food for thought for us, clinicians who, dealing every day with mental suffering may lose the contextual sense of the transformations we promote.

Blind loyalty to any psychoanalytic current has never been either a guarantee of validity or the only means through which the potentiality of thinkability can be expanded.

All my work is based on the assumption that it is possible to refer to transformational models in analysis; all cognitive, emotional and communicative processes are based on a specifically informational approach: Transformations are narrative evolutions of transcoding (for example from preverbal to symbolic) decoding (from unconscious to conscious) recoding (for example the reintrojection of a projective identification developed by the analyst) met coding (for example the formulation of general theories on the base of data and observations) of different ideas and information constituting our knowledge , conscious and unconscious, correct or distorted of the world and of our own representations (concepts, affects, beliefs, behaviors).

According to cognitivist psychology they are data referring to different historical and cultural contexts, collective and individual, all highly subjective.

Affects, in particular, "promote learning as they constitute an important signaling system enabling organism to acquire new patterns and to avoid others; they are an important reinforcement of motivation and stimulate the activation of action plans. Moreover affects contribute to solve problems, activate memory patterns, and affect perception. To sum up, emotions indicate the contribution of the paleoencephalon to the organization of signals in messages which prepare to organized actions and therefore to knowledge (Fossi 1991).

Transformations are possible because complex interactions are ruled by the law of necessary variability, according to which the range of variations a system can cope with without disorganizing is proportional to the range of variations inside the system. In other words the more constraints there are in a system the more degrees of freedom exist and the more degrees of freedom there are the more the system can evolve, generating emerging qualities of knowledge objects and tools of knowledge.

It is not possible to establish, as von Foerster states, whether the world is an object to be represented or it is a construction of ours; we can only consider knowledge instantaneous, emerging from the coupling of the transformational systems and environments.

This approach can be defined post-constructivist as it gives dignity to what is transformed, which must exist to be object of transformation. It grants also naivety to the observer, who is certainly more aware and able to leave hackneyed ways of thinking behind.

Free from all that ,we can therefore consider organization as knowledge and mnestic trace of itself, which goes beyond the principle of entropy and gives creative evolution the gift of necessary variability which destroys to create and creates to destroy.

There can be indeed only an imperfect knowledge, temporary and by definition open to infinite possibilities.

Applying a complex perspective means leaving aside the previous theoretical references to give space to new points of views and new approaches "we know too much to understand something new (...) it is necessary to go beyond this hypertrophy of awareness which prevent from any possible change" (Ferruta Galli 1992).

According to Gargani, it is necessary to arouse interest in new events without analyzing them through old perspectives; it is necessary to wonder through the new to abandon the possible functional inelasticity which prevents the expansion of thinkability (Gargani 1988).

Complex perspective referring to the comprehension of phenomena can lead to more flexibility in the conception of the various psychoanalytical approaches and permit if not a complete integration at least a enriching dialogue, with no preclusions and prejudices.

There is no opposition between apparently incompatible concepts, such as drive and relationship but a dialogic, antinomic relationship in a unique emerging phenomenon.

Welcoming a complex perspective implies accepting an inclusive logic, an "and-and logic" rather than exclusive "or-or", logic, a way of thinking open to the evolution of concepts, not blocked by the limits of the common logic.

We are taken therefore into a permanent laboratory which does not bar possibilities to question the fundamentals every time we do research or deal with clinics, free to use the accepted paradigms but also to update them based on new discoveries which can require new context or new tools.

The clinician becomes an epistemologist in order not to bar possibilities of development of their discourse, the epistemologist carefully observes the clinician because new paradigms can be produced on the basis of their receptivity.

Science, as Kuhn reminds us, is not cumulative, but consists of paradigmatic revolutions, with a relationship which is often inclusive and psychoanalysis, perhaps due to its own nature, experiences continuously this kind of revolutions without necessarily disproving previous discoveries.

It is not unusual to find Freud's contributions in present theorists, who through different languages reach the same conclusions, or to read them from the modern epistemological point of you without finding any incongruence.

Even if complex perspective reasons on paradigms, it does not imply a change of paradigms.

Khun means, by that word, “scientific achievements, universally recognized, which for a certain period, provide a model of problems and solutions accepted by researchers in a certain field.”

This is not the matter, since to expand thinkability does not imply choosing a certain metapsychological or clinical approach (Khun, 1962).

To get involved in analytic research on this basis means not to take for grant any acquired tool, it means not to settle for an aspect of our discipline in order not to limit the acceptable solutions whether they are assertions or theories.

Chapter 1

AT THE BEGINNING

Clinicians tend to have an aversion to theoretical reasoning and when it does not originate from the analytical relationship their impatience may turn into real suspicion.

The fact is that "we need a theoretical apparatus for any empirical research", since truth is quite often so improbable "as to require not only artistic/poetic sensitivity and audacity but also an observational rigor sufficient to convey a meaning shared and accepted by the scientific community" (Stafford Beer 1992) (Bodei 2000)

Psychoanalysis has, in fact, all the more to follow a path which is opposite to that followed by many other disciplines. From the beginning it was expected to study its own object under conditions which could not neglect the relational context to avoid that object's dissolution. Such a question, due to its complexity, risks wavering in a potential interpretative "Babel".

Psychoanalysis is thus allowed to use, to a far lesser extent compared with other sciences, that heuristic simplification so useful to define properties and laws, which puts the observed object in non-complex, experimental situations divorced from their relevant context.

Furthermore, "psychoanalytical science, dealing with premises, risks becoming religious doctrine more than natural sciences, as it lacks the criteria of measurability and objectivities, the protection assured by magic pseudo-scientificity of number" (Cremerius 1985).

Also the recent hermeneutic tradition maintains that knowledge objects are dependent on historical heritage: Hans George Gadamer, Heidegger's pupil, suggests that speculative truths are so affected by preconceptions and prejudices of those who investigate truth that, orienteering differently our preconceptions, we can get new interpretations and reconsiderations of knowledge objects.

Therefore cognitive experience has got a historicity which influences the quality of comprehension.

The relationship with knowledge objects is neither subjectively aggregated nor aseptically objective, but the interpretative action consists in an adjustment mediated between our interpretative coordinates and the ones inherited from the past.

It follows that history itself is a comprehension which can undergo new interpretations and further comprehensions based on present changes in the structure of preconceptions.

While the philosopher still maintains the dichotomy between spiritual sciences, subjected to the relativity of history, and natural sciences pursuing an ideal of objectification, after

Feyerabend we cannot but agree that even the latter are subject to the same relativizations and this implies their greater heuristic affinity.

This reasoning makes natural sciences coincide with human sciences at least in a broad sense.

Also knowing them in fact implies an interpretative background which cannot be disregarded.

Res cogitans and res extensa are part of the same “language” , of an hermeneutic matrix which permits us, as psychoanalytic researchers, not to divide to conquer language but on the contrary to unify the different paths of knowledge to fully understand what communication and signification can be.

In terms of hermeneutics therefore, the epistemological peculiarity of analytic method would be an investigation how rational subjectivity records, suffers, understands and expresses the forms of its dependence on drives of the biological body. The language being investigated is actually, not only the symbolic language but also the body language relating to symptoms, the concrete language relating to actions , the cultured language relating to conceptual expression: that is to say all the combinations you can find along an axis going from the somatic to the mental , the cultural and vice versa. Psychoanalytic semiotics recognizes therefore a continuum going from corporeality to verbal language. (Barnà 1992)

As Gianni Vittimo reminds us, commenting on Gadamer’s theories, the method adopted by natural sciences, that is to say “the means by which, a subject originally conceived as if it was opposite its object, ensure the availability of the latter” has been taken from the spiritual sciences, *Geisteswissenschaften*, demonstrating hence its inability to understand the truth. This however caused them to be relegated them to pure aesthetic experiences having nothing to do with discovering the truth, the true and the false.

The world of beauty is separated from the world of truth, the world of appearances is opposed to the world of reality; this is manifested in Kantian criticism and got its recognition in Shiller.

Splitting is thus praised, imagining “an autonomous world beside the real world and with no relation to it”; aesthetic awareness is born.

“Aesthetic awareness (.........) is not the result of a misinterpretation which experience gives of itself in light of modern prejudice charging sciences and their method with the knowledge of the truth”.

Aesthetic experience, actually changes its author, a work of art is able to reveal new things in its creator, in this sense it is a means of self-comprehension.

Through a work of art a transmutation is carried out , disclosing nothing but the representation of what is represented, that is to say the truth of phenomenological experience.

Art is therefore an extra-methodical function of truth, an interpretative and executive practice of truth, it becomes an object of investigation of a world far from the Kantian idea of truth, feels the experience as self-evident: here truth means also “objectivity, but only in the sense of the effectiveness of a meeting with the other, another who understands me better then I understand him” (Vattimo1965).

Gadamer reminds us that according to Plato the essence of thought is the inner dialogue of the spirit with itself, a linguistic process of thought which develops in and through language.

It is in the inner conversation that the world begins to open up and to form in every field of experience.

Natural sciences were able to take for granted that the *simple* exists, actually, quoting the philosopher Bachelard, in science nothing exist but the *simplified* (Bachelard).

And simplification selects nothing but what "is somehow interesting to the investigator and removes all that is irrelevant to his goals "it calculates the stable, the given, the certain and avoids the uncertain and the ambiguous, it produces a kind of knowledge which can be easily processed for action and by action" (Morin, 1986)

As far as the object of psychoanalysis is concerned, there is not a thinking that is not intersubjectivity and relationality since "the mind is rooted in history, shaped by our experiences, inhabited by thousands of speakers , on different levels more or less conscious" (Tobler 2000).

Therefore it is reasonable to agree with George Lukacs about a theoretical approach which conceives "the complex (....) as the existing primary element and consequently we must analyze the complex as complex and then move from the complex to its elementary elements and processes" (Lukacs)

Freud was the first to face this problem: eighteenth-century positivism's son, he was not able easily to do without the deterministic concept, according to which the world is a simple machine whose working is theoretically predictable even in its most peculiar or ?

Especially early Freud was seduced by the great theoretical principles, on which that conception was based, that "paradigm of simplification"; principles which were started by the great Cartesian reflection in the spirit of the disjunction between the thinking thing (ego cogitans) and extended thing (res extensa).

He was first seduced .by the principle of *disjunction* ensuring the definition of research objects, their delimitation, their quantification, in short their emerging from the blurry background of a very untidy and little unwieldy whole. This principle then distinguished the areas of competence of the various branches of science, hence the peculiarity and the independence of physics from chemistry, of chemistry from biology, of biology from psychology.

By this division we come to know, the more I divide the better I know. This has proved a great principle, as it accompanied the development of natural sciences over their history.

The principle of disjunction is paradoxically complementary to the principle of reduction, hence if I want to know I must reduce every level of reality to the previous one which is supposed to be simpler.

Biology must be neatly reduced to chemistry, that is to say it must be explained, according to the laws which bond molecules together, in the same way psychology must be reduced to biology.

In the period up to "*A Project for a Scientific Psychology*" (1895) Freud was busy trying to demonstrate this principle, trying to explain in neurological terms what he was realizing on a psychological level in the field of neurosis.

And certainly he was not either the only one or the first one to make this attempt , one only needs to think of the extensive work by Sigmund Exner, the physiologist who was Bruke's successor and had discussed about that in *Entuwwrf zu einer physiologischen Erklarung der psychischen Erschinungen (Draft for a physiological explanation of mental phenomena* , (Wien 1894)

During the same period Fechner tried through his psychophysical research to explain the relation of causality between psychic phenomena and physical phenomena in mathematical terms taking their reciprocal reducibility for granted.

Freud on his part in his Project was going to set up a well-grounded mechanist theory which was bound to last.

"The goal of this project is to offer a psychology conceived as a natural science, that is to say to describe psychic processes as quantitatively determined states of material particles which can be identified in order to make them clear and incontrovertible. (Freud 1895)

Although this was a real manifesto of the principles of disjunction and reduction suitable to the most radical paradigm of simplification, the complexity of the nature of his object of investigation must have led him to a more careful and sometimes hesitant behavior.

If we examine his correspondence with Fliess from 1895 and the following year, we notice some enthusiasm in the letters of June and August, then Freud seems discouraged some days later; in October he recovers until November, when he sinks definitively into dejection, giving up on his *Project.*

He never resumed it, but he kept many reductionist ideas till his last theory about drives dualism, as "*Beyond the Pleasure Principle*" (1920) demonstrates, and he never changed them.

Formulations of mechanicism certainly lost the univocality by which they had been conceived according to which dynamic and economic concepts took into bigger account the different nature of physical forces compared to the psychical ones, but his secret faith in his Project was still evident in his 1920's work; he still maintained the principles about "quantity of energy" and "parts of the central nervous system". (Freud 1920)

Even if he abandoned the biological model of the Project to arrive at reasoning about subjectivity, Freud keeps his nosological reformulations within the traditional Galenic-empirical model adapting them to its principles (Bordi 1989).

Therefore fixation of libido is aimed to keep somehow a certain link with biological determinism in the origin of symptoms, as libido is actually a "quantitative characteristic, even if it is not measurable yet", on the border between psychic and somatic functions (Freud 1921).

Pathogenesis remains specific for every "nosographical entity" like "therapeutic specificity" is connected with clear clinical presentations, so there is a specific therapy for every clinical presentation; childhood trauma equally corresponds to *noxa morbosa.*

A further cornerstone of simplification consists in abstracting an operating rule of a given phenomenon from the chaotic weave of events happening simultaneously, in order to identify the supposed supreme principle which rules things.

Freud tried bravely to praise this cornerstone too: in *Further Remarks On the Neuropsychosis of Defense* (1896) the most radical observations by Freud as a biologist are only a memory but the principle of abstraction is very popular, nothing seems to weaken the passion for simplification, now that the psychological ground has shifted.

In this work the symptoms of obsessive neurosis, hysteria and chronic paranoia, are all attributed to a similar etiology concerning childhood sexual trauma; this is a remarkable novelty, as the genesis of the pathology is identified in a combination of psychological factors.

Furthermore, in this work he proposes a conflict between instinct and its repression which will have a leading role in all the next theories (Freud 1896).

Therefore among all factors affecting psychic life, childhood sexual experiences seem to be the specific cause of the variety of symptoms arising in adulthood.

This work is contemporary with another one, *Heredity and the Aetiology of the Neuroses* (1896) which paves the way for a psychological conversion against reductionism.

It is enough to notice that in this work the factor “heredity” (biological component) is considered, just a condition for pathogenesis, clearly separate from the specific causes (premature sexual ones) and concurrent causes.

In this regard, the following passage is quite revealing; in the metaphor the author uses, we can notice also his diminishing nostalgia for objective determinism.

“The action of heredity can be therefore compared to the action of a multiplier in an electrical circuit which increases the visible movement of the nail but cannot determine its direction” (Freud 1895).

Freud actually gives even less importance to the contributing causes relegating them to the category of “triggering factors”, as we nowadays call them, even if he maintains that he does not want to “undervalue the etiological importance of these common agents” (Freud 1895).

One could suspect that Freud at this point was left hanging in the balance, to one side leaning on biological positivism and to the other lolling in an unknown and confusing universe, reluctant to get bridled by the coordinates of simplification.

Division seems dangerous but necessary, reduction seems irrelevant and very complicated, and abstraction may be a good way out.

But actually it all must be redone, it just doesn’t add up.

Freud, in a letter to Fliess in September 1897, in gloomy despondence, reconsiders, in the light of his self-analysis, the idea that specific etiology refers to memories of sexual abuse which really took place, now he thinks that memories are nothing but fantasies developed later

The clinical case of Miss Elisabeth von R., included in *Studies on Hysteria* (1892-1895), shows how inner processes following a trauma are pliable hence the mind’s functional ability to rebuild itself continuously in a complex, uninterrupted weave of experiences, perceptions and distortions.

In fact there begins a shift from the external world to the causality of the inner world; psychoanalysis is born in the true sense of the word and the search for the real causes of neurosis is restarted.

But above all, Freud can now look out into the boundless universe of psychic complexity which seems to require various levels of structure which intersect in a multitude of causes displaced throughout a varied temporal order.

The time variable deepens the mystery and renews the challenge, suggesting levels of causal imbrication which are unprecedented for late 18th century science.

Time, that strange factor, which had kept philosophers' reflection engaged until it became a real puzzle, which had flirted even with theologians until exhaustion, knocked at the back-door, asking to be admitted again to the assembly of scientific thoughts.

The efforts aimed at talking it down were useless; mechanist physics had tried to do that, reducing it to a simple constant but now also from that side of human knowledge time was seeking justice.

Chapter 2

MYTHS DISPELLED

It is this way that Albert Einstein starts demolishing certainties precisely about the inelasticity of time, binding its destiny to the interaction of other physical variables and reviving the real features of this concept from his point of view of the world.

Einstein and Freud need, considering the complexity of their respective fields of study, the same epistemological approach.

They both study phenomenal fields where it is necessary to apply axioms which organize the field of the observed phenomena preventively in order to achieve the construction of models.

That is because there is a very big gap between the phenomena being observed and their description and theoretical explanation, unlike what happens in other fields, such as physical mechanics.

In other words such a gap is filled in with theories which seem to derive from the observed phenomena but actually precede them.

This separates Freud even more from the naive positivism from where he seems to have started, "Comte himself claimed that without a theory through which phenomena can be connected with principles, not only would you be unable to combine isolated observations and reach useful conclusions, but you could not even remember them and in most of cases even observe the facts" (Chianese 1988).

To sum up, it seems so impossible to fill in the logical gap between experience and theory that "the most abstract levels of metapsychology, like similar levels of any other scientific theory, can neither be proved nor disproved by any factual evidence"

In order to explain this gap both Freud and Einstein use intuitive connections which are really far from their fields of study (Napolitano, 1988). Chianese underlines how the two theoretical fields support and use each other to explain empirical facts. Einstein bases himself on psychology, maintaining that axioms are psychologically grounded in experience, as there is no logical path leading to experience but intuition; Freud, for his part, justifies his own theoretical conclusions epistemologically, claiming that he proposes them in the same way as physics does!

The strictly epistemological approach seems not to be able to grasp the concept of invention referring to an indefinable *quid.*

Einstein's position about the construction of scientific thought is even more radical than Freud's.

He thinks that imaginative freedom is useful not only to define theory but also to imbue a general dimension of thought.

Einstein, like Freud, thinks data are not unchangeable observable entities and concepts deriving from them can be reconsidered and reexamined as they are not a priori concepts.

Influenced by Hume's and Mach's critical analysis , Einstein reconsiders the notions of space and time taking them away from the absolute realm where philosophers had confined them, but contrary to their strict empiricism, he thinks that constitutive concepts of knowledge are free creations of thought obtained inductively through sensorial experience.

That is because there is an unfillable gap between concepts and sensorial data.

In a sense he draws on Plato but also the more recent anarchist theories of contemporary epistemology; furthermore, concepts are still constructions of mind like the notion of the whole, which although it dates back tout court to the pre-scientific era of philosophy, is still usefully applied.

Hence according to Einstein atoms they are conceptual objects; it is still reason which builds a structure of knowledge, even if Einstein remains faithful to the idea that thought is somehow able to grasp reality and its functioning.

He is firmly convinced that it is possible to find the deep harmony which is inherent in the order of the real and thus he gives credit to Leibniz's historic theories but also explicitly to Max Planck's (Moszkowszi 1921).

In some way speculation fills in the gap between reality and knowledge by means of mathematical construction, with its simplicity and elegance (Einstein 1931, 1933).

After all Einstein's theory of knowledge proves to be a correspondence theory, opposed to Bertrand Russell's , but also to a metaphysical escape into a concept of the world which denies the existence of an object outside of the cognitive apparatus.

The object does exist, it is independent and analyzable.

He is a strict empiricist especially when he speaks about the axiomatic construction of Euclidean geometry, based on the correspondence between thing and concept.

To hypostatize concepts can be useful if their link with the empirical is not disregarded.

Here Einstein is deeply anti-Kantian, stating that there are no inalterable and a priori concepts.

Axioms can be chosen freely but their validity is assured by their comparison with experience, even if mathematical construction remains an absolutely creative principle of theoretical physics.

He did not give up the atomistic theory; on the contrary, he applied Boltzmann's probabilistic theory about entropy to the study of radiation, which allowed him to use the concept of quantum of radiant energy. This atomistic analysis finds a correspondence in Bohr's theory, according to which atoms can transit from a steady state to another one with quantum jumps. Even though mathematics and experience are *complementary functions* in the knowledge construction, the first has the freedom that not even Kant's pure *intuition* could offer.

This was influenced by *a priori forms of intuition,* which is not valid for Einstein because he accepts the discoveries of non-Euclidean geometry. He thus gets closer to the Platonic concept of *mathematical objects* (Einstein 1921) (Lenzen 1958).

Einstein was as interested in knowledge theory as he was in the basic principles of physics, which is quite rare among scientists and makes him an influential representative of modern epistemology.

He does not choose Bridgman's operational point of view, on the contrary his approach is totally unrelated to naive realism. His reasoning about foundations of knowledge makes him more similar to a Greek empirical rationalist rather than an English empiricist like Hume.

The debate between the conceptual world and experience is a fertile exchange of which Galileo was the first proponent. According to Einstein not only the logical method of formal implication is ineffective but so is the one of extensive abstraction developed by Aristotle and Whitehead; scientific knowledge can aim for nothing but an ideal meeting between thing and concept. (Einstein 1933).

We can deduce that Einstein was not a positivist at all, theoretical results do not directly follow empirical data and on the contrary theoretically based knowledge reveals the thing itself, which can be investigated through scientific means.

However the Theorist of Relativity still considers the principle of simple causality valid even if he distances himself from Hume's point of view. He remains convinced of the principle of determination according to which it is always possible to foresee the future state of a system (Northrop 1958).

Einstein suspected the limits of positivist approach, which is demonstrated by his progressive moving away from Ernst Bach's epistemological theories. In Mach's obituary which the physicist wrote in 1916 we read: "science is nothing but comparing and ordering our observations through methods and points of view which we learn by trial and error (....) This ordering activity results in abstract concepts and rules connecting them (....) Concepts make sense only if we can mention the objects to which they refer and the rules through which they are referred to those objects (...). He (Mach) thought that every science had the task of ordering the single elementary observations which he described as sensations" (Frank 1958).Einstein accepted all that, even if in this programmatic statement he saw an extremely simplistic approach, which did not take into consideration the level of abstraction of theories like the Relativity theory itself.

In essence, fundamental principles cannot be verified directly through sense experience but the quantity of logic relations is so large that it is difficult to compare the abstraction of every fundamental principle with experience. His reasoning is actually more similar to Carnap's logical empiricism, according to which affirmative clauses are based on affirmative implications which only secondly prove to be comparable with the perceptive foundation of experimental observations (Carnap 1939).

Einstein, like Freud, seems to be closer to the positivist manifesto by Comte than to Mach's programmatic statement.

In an article on positivist philosophy Comte already wrote: "If on one hand every positivist theory must be based on observations, on the other our mind needs a theory to make observations".

Einstein surprisingly claimed that foundations of scientific theories are in psychological connections and not in logical ones. In other words, only by creating conceptual apparatus beyond logics can one, through images, signs, metaphors, grasp the ultimate reality who wants to conceptualize.

There is an explicative quantum, which is for Einstein and Freud always beyond experience, this quantum is in a sense *metaphysical* (Of course I use this word here to underline the immateriality of this concept), not separate from experience but on the contrary mysteriously and closely related.

Also this feature distinguishes Einstein's and Freud's positions from Mach's naive positivism and opens up new really innovative perspectives.

It is the epistemological jump that behaviorism and psycho physics have never done, missing the chance of complex explorations of the psyche.

The logical-empirical position reveal spheres of thought remaining anchored in experimental data. Here Einstein demonstrates his depth as a scientist; his thought restores the role of mathematical language at the service of natural knowledge beyond empirical data, which are classifier, quantifier and sclerotic due to naive empiricism and opens up to new dimensions of knowledge.

Speculative optimism by which he is motivated rejects the abstract mathematical concepts developed by Henry Poincarè, denying an "exact base" to science on the whole.

Einstein's and Freud's contributions are two important previews of the modern study of nature, even if their vocation as scientists consists in searching for simple and general laws able to explain very broad fields of natural reality.

According to the Physicist, the general theory of relativity was to be the basis of a theory concerning forces acting in the universe, according to the Father of Psychoanalysis the study of a single case would provide information which could be generalized thanks to the conflictual structure of constellations which were nosologically similar.

In this sense they are still pre-complexity scientists, unlike some of their contemporaries who were starting to go beyond these theoretical approaches.

Bridgman already wrote in 1927: "Simple means for us what seems to be simple from a conceptual point of view (....). It is clear that our thought must follow the lines imposed by nature on the mechanism of thought itself: but how can all our nature accept these limitations? (....). Well, our reasoning has explained something, that is to say that our concepts are not well-defined at all, on the contrary they are fuzzy and cannot be applied to nature, a lot of them can but approximately, only to a limited part of it (...). The strange thing is, rather, that there are a lot of apparently simple laws.

This remark must be repeated for all simple laws of nature formulated till now; they can be applied only to a restricted area (...). It will not seem strange, therefore, that in a restricted area, where the most important phenomena are of the same kind, nature's behavior can follow relatively simple laws" (Bridgman 1927).

Philip Frank insists, however, on claiming that Einstein and Bridgman's positions differ actually for the emphasis placed on one hand on the charm of simplicity on the other on the attraction of nature's complexity.

Einstein took the vision of nature a further step forward, according to a concept of knowledge which anticipated the theory of complexity; he considered the concept of space and time as if it were no longer revealed by the world of ideas, or created by pure reason, as philosophical apriorism maintained purported to establish

Complexity proposes the correspondence perspective which is not aprioristic, not founded on ontological or metaphysical bases and not simplified; it is a theory of knowledge composed of observers who observe observers, a theory which is definitely non subjective.

"Speaking of different observers is only a means to describe the plurality of conventional systems (...) Relativity however does not mean abandoning truth; it only means that truth can be stated in different ways (...). All relativities in Einstein's theory are of this kind (...). His relativity is not conventional as the conventionalist overlooks the fact that only the enunciation (...) in which every reference to the definition of incongruence is omitted is arbitrary; if the enunciated is completed by adding this reference, it becomes empirically verifiable and thus acquires physical substance" (Reichenbach 1958).

That is diametrically opposed to logical Kantian apriorism, the field of pure reason gave way to an edifice of mathematical assertions dissolving the *Synthetic A priori* and finally building a system of relations and statements which are abstract but experimentally verifiable in a phenomenon.

Similar to what Locke thought, there are no innate ideas, but only acquired ideas, even if "the axiomatic basis of theoretical physics cannot be extracted from experience but is a free creation of the observer's mind" (Einstein 1933).

As Wittgenstein and Feyerabend explained later, cognitive processes consist of practices which are socially, ideologically and aesthetically conditioned and which are, as Einstein seems to warn, at the base of the influence of ideas in the development of any scientific theory.

Therefore it is useful to set ideas free in order to go beyond functional fixedness which blocks thought all the more if one leaves strictly factual reality to explore phenomena which require a high level of abstraction.

"A scientific theory consists of a formal axiomatic system; it is therefore built through a language. It contains in fact individual variables, logical and extra-logical constants which play the role of invariable elements defining what is called grammar in natural languages; moreover it is provided with a syntax that is to say with a system of rules which determine the formation and transformation of the constituent parts of calculation.

Now, both using and applying this language in research is a practice, as it is a collection of choices of constructive actions according to logical-mathematical criteria , *which do not necessarily follow a rule, but possibly create it , becoming regulatory. All that sounds like a paradox but actually it is not, because the tension of the conflictual but also co-operative relationship between rule and deviation, is the creative condition through which knowledge proceeds*" (Morlotti 2002) (emphasis added).

Freud still highlights the validity of empirical verification which does not necessarily lead to the verification of a theory (Freud 1915) but at least enables the connection of the observed facts through operative definitions which keep the theory alive.

Both Einstein and Freud hope for and seek a simple model of constructions underlying the realities to study and this is still affected by a speculative optimism which is not completely immune from the temptations of the paradigm of simplification; it is nonetheless possible to understand that the complexity of the phenomena to study requires a continuous epistemological revision which Freud certainly never interrupted.

The reflections to which Freud found himself constrained were in good company, they concerned the whole *Weltanschauung* of the science of that period, as they affected also physics, a prominent science, true and honest, which was not disposed to admit it had demeaned itself by its own hands.

He was moving from a simple model, where only the past affects the present, to a non-linear determinism, where the present and the future can reprint memories and therefore have even more pathogenic effects.

The concept of psychic abreaction gives way to the concept of *Nachtraglichkeit* or afterwardness, which Jacques Lacan called *après coup* in French, meaning that "the present material of memories is periodically *retranslated* according to new relations, it is someway *reprinted*" (Laplanche 1993).

Freud arrived at the concept of *overdetermination,* which can also be called multicausality or *multiple causality;* it is a concept in common with probabilistic physics, according to which a given cause is not necessarily followed by a single effect but it can produce a range of more or less probable events.

If it seems a paradox, an absurdity, an antinomy, it is only because of our difficulty to understand the unpredictable nature of effects in the interaction of different variables and also because of our resistance to conceive those effects as partially random.

The relation of causality is considered exhaustive and self-sufficient, perhaps due to the innate tendency of our cognitive to simplify.

When one deals with psychic phenomena it seems difficult to defend the bulwark of the principle of abstraction, therefore it is necessary to admit that the whole paradigm of simplification is totally insufficient to include new discoveries both at theoretical or empirical level.

In *The Interpretation of Dreams* (1899) Freud starts to explore the field of the psychic unknown disregarding epistemological matters, the heritage of Positivism (Freud 1899).

Particularly, with the concept of the unconscious as a further dimension of personality the problem of causality as a mechanistic matter gives way to the unexplored aspects of the nature of the unconscious.

Here the principles of Aristotelian logic are not valid, neither the principle of identity, nor the one of non-contradiction; neither the principle of the excluded middle nor the one of sufficient reason; even the matter of time changes its nature, it seems indeed that time does not exist in the unconscious , the past and the present are equivalent.

In these conditions it is impossible to conceive something like a causal explanation as is generally assumed in the scientific debate.

It is true that in 1901 Freud still believed that the principle of causality concerned, rightfully, the inner events of psychic life and he remained convinced of that for the rest of his life, but it is reasonable to suppose that this concept was more accepted empirically than defined theoretically.

In *Fragment of an Analysis of a Case of Hysteria* (1901) the criteria of disjunction and causality are still interwoven, even if, since then, the matter of the interaction of the factors involved in psychism remained unresolved because Freud gradually realized how complex his own object of study was (Freud 1901).

In this work, in fact, "the live report of the analytical dialogue seems to be characterized, as far as the methodological and linguistic structure is concerned, by an interrogative style, where the *object (disjunction)* is browbeaten and is expected to confess in a progressive and forced revelation of the truth (*causality*).

The specific metaphor proposed by the author himself is the one of the "doctor dissecting" states of mind.

At the same time, mostly in the last part of the work, an *attitude of waiting, of peculiar receptivity and suspended judgment* seems to appear: an almost amazed or perhaps admiring overview on the extreme complexity of the inner world of the patient" (emphasis in brackets added) (Melandri 1989).

The dissection of consciences reveals a complicated knowledge system where thing representations and word representations are elements typical of every level of representation, either unconscious, preconscious or conscious, and interact within every level in a horizontal and biunivocal sense (for example unconscious-unconscious) and in vertical and biunivocal sense (for example unconscious-conscious) as happens as a result of an interpretive action.

Freud was aware of the interdependence between the levels of representation; in 1915 with regard to communication between the conscious and unconscious system he makes this assumption explicit: "It would be wrong to imagine that the *Unconscious* remains inactive and that the *Preconscious* does all the psychic work, that the *unconscious* means nothing, that it is a rudimentary organ, a residual of the evolutionary process. Or to suppose that the relationship between the two systems consists only in repression, where the *Preconscious* throws all that seems disturbing into the *Unconscious* chasm. On the contrary the *unconscious* is living, able to evolve, and it has other relationships with the *Preconscious,* including collaboration.

In short, we must say that the *Unconscious* extends in offshoots, that it lets itself be conditioned by the events of life, that it constantly influences the *Preconscious* and that it is even, in turn, influenced by the latter.

If we expect that the two psychic systems are precisely and strictly divided, the study of the *Unconscious* will disappoint our expectations" (Freud 1915).

Freud, however, brings the conscious and preconscious much closer to each other than to the unconscious and evidential reasoning comes, as usual, from clinics: "On one hand we have evidence that even an intellectual work, subtle and difficult, *(applying the principles of Aristotelian logic)* which generally requires exact reasoning, can be done in a preconscious way without reaching consciousness. There is no doubt on cases of this kind: they occur, for example, in sleep.

A person, just after waking up, can get the solution to a difficult problem, of mathematical or other kind, which they unsuccessfully tried to solve during the day" (Freud 1922).

The process of destructuration carried out by an interpretant word results in new connections between the representative elements at every level of the knowledge system, in a circular transforming dynamic (Cargnelutti Muratori 1990).

The psychic system can be considered a complex knowledge system, which is intrinsically congruent at every representation level, provided with its own methods at every different level; at the unconscious level with the principles of contradiction, included middle and indeterminism; at the conscious level with the principles of non-contradiction, excluded middle and determinism.

This system seen as a whole would consist of knowledge subsystems provided with conflicting modalities, notwithstanding this without a logical apparatus, just because it is based on the transcribability of the respective contents, which is recognizable also empirically.

In the aforementioned work of 1920, *Beyond the Pleasure Principle,* Freud resumed the discussion on psychic causality, this time from the point of view of metapsychology, analyzing it through the concept of repetition compulsion, which was first dealt with in *Remembering, Repeating and Working Through* (Freud 1914) and relating it to the great conflict between the death and life drives.

However Freud's attempt is too speculative; he examines the reason of compulsion basing himself on abstract concepts which lead to unsolved contradictions, regarding the undecidability of which drive is dominant.

Finally, in *Inhibition, Symptoms and Anxiety* (1926) repetition compulsion is considered a resistance, a typical process of the unconscious function but the enigma remains unsolved. It is simply bypassed, avoiding the hypothesis of confliction between life and death drives (Freud 1926).

Freud's final metapsychological reflection is still today a most obscure point.

It provides at most the basis for metapsychological postulates with different approaches, but nothing verifiable emerges.

We should start from this same trail of thought without expecting to understand everything, otherwise we would fall into the trap of naïve positivism, instead aiming to understand something closer to the psychic reality, with the multiple interactions of the factors involved, with uncertainties and with chance.

Chapter 3

EVOLUTION AND NONLINEARITY

What is the relation to a world like that? Well, we are made of the same elemental compositions. So we are part of this thing that is never changing and always changing. If you think that you are a steam boat and can go up the river, you are kidding yourself. Actually, you are the captain of a paper boat drifting down the river. If you try to resist, you're not going to get anywhere. On the other hand, if you quietly observe the flow, realizing you are part of it, realizing the flow is ever changing and always leads to new complexities then every so often you can stick an oar into the river and punt yourself from one eddy to another.

Brian Arthur

One of the most innovative features of Freud's theory is certainly its openness to accept the plausibility of concepts which are apparently contradictory.

Freud shared this characteristic with other great thinkers and innovators, such as Newton, who already in his work *Opticks* (1704), explained the motion of light using both particle theory and wave theory which are apparently in contradiction.

As far as Freud is concerned, an example of this openness is found in his discussion of the concept of complementary series which he introduced, as known, in the 1915 edition of *Three Essays on the Theory of Sexuality* and which resumed and developed in *Introduction to Psychoanalysis* (1916/17*) and in Analysis Terminable and Interminable* of 1937.

In the latter he solves an apparent contradiction admitting that both heredity and environmental factors can be implied in defining the quality of the ego. He says: "The qualities of the ego which we feel as resistances can be caused by hereditary factors or equally be developed during defensive processes" (Freud 1937).

Discussing complementary series he is even more explicit, claiming that both genetic and environmental factors are decisive for the etiology of mental illness.

Therefore, illness is caused, as Charcot thought, neither only by heredity, nor completely by environment, but by the complex imbrication of these two factors. Even if in the same work Freud sometimes insists on the genetic factors of pathogenesis, as one can infer from this unmistakable passage: It is not due to a mystical overvaluation of heredity if we consider reliable the hypothesis that for a no-existing ego , the directions of its development, the tendencies and the reactions it will show later are pre-determined (...) our analytical experience persuaded us that even some really specific psychic contents as symbolism originated exclusively from hereditary transmission; moreover, based on various research on folk psychology, we should suppose that other equally specific sediments of the ancient human evolution are part of the primitive heredity" (Freud 1937). Freud was actually not the first to become aware of the causal imbrication in the natural field; he was certainly preceded by one of his greatest inspirations, Goethe, who was well-known not only as a poet, but also as a scientist and natural philosopher. He was appreciated more as a poet and his modernity as a scientist was recognized later in contributions from outstanding personalities such as Darwin and Freud.

In Goethe it is already possible to find the proto-Darwinian idea that man is the result of a slow creative natural process and that its characteristics do not distinguish him completely from the rest of the animal kingdom. He never conceived an evolutionist theory, but he perhaps contributed to instill doubt in the scientific thinking that a plurality of causes possibly originated in a complexifying journey which is common to all natural species.

Freud first approached natural science when Darwin's theory of evolution through natural selection was at its greatest splendor.

During the period of his university studies he was a student of Theodore Meynert's, outstanding psychiatrist of the time, author of the work *Psychiatry* published in 1884; this work referred to *Expression of Emotions in Man and Animals (1872)* by Darwin which was quoted twice in *Studies on Hysteria* (1892-1825), Freud's first work of psychology written in collaboration with Josef Breuer.

In the above-mentioned work, Darwin meant the physiognomic expression of emotions as functional to survival and thought that the least expressive behavior, even if it seemed to have no meaning at that time, should have been meaningful at least for the previous human species.

Freud is in such complete accordance on that point that he supposes that also neurotic symptoms cannot be casual expressions but must have a precise meaning which is in some way useful.

Meyner, on his part, considered Charles Bell and Charles Darwin the first naturalists who ascribed an expressive importance even to breath rhythm and Freud resumed this idea in a work of 1895, *A Reply to Criticism of my Paper on Anxiety Neurosis,* relating breath to sexual excitement (Freud 1895).

Expression of Emotions in Man and Animals deeply influenced Freud's thought, according to which Darwin gave a triple explanation of expression of emotions.

The first principle is what is called useful associated habits and it consists in the above-mentioned theory according to which every expression of emotion once must have been useful to survival, as the attack posture of a dog is functional to the preparation of the physical attack against another dog or any other aggressor.

The second principle claims that some expressions which do not seem functional can organize themselves antithetically in order to express contradictory emotions.

Freud did not use much of the second principle in his works; on the contrary he used abundantly the third principle, also known as Herbert Spencer's general law according to which an emotion exceeding a given threshold in an animal can be manifested thanks to an excitement of the innervation which stimulates its musculature.

Freud based himself on this law to formulate the constancy principle which is the foundation of the economic theory of psychoanalysis.

It is no coincidence that it already exists in its first theorizations as a regulator of the functioning of mental apparatus in order to explain how the psyche tries to make its level of excitement as constant as possible; this constancy can be kept thanks to a mechanism of avoiding, discharging and defending of extra energy. The constancy principle is strictly related to the pleasure principle, as pleasure would correspond to a decrease in energetic tension.

Later Freud realized that the relationship between pleasure and pain should have been much more complex than what he had hypothesized but the initial temptation to relate the study of psychology to the general laws of physics was so strong that he was not able to consider the new discipline a real science.

The constancy principle resembles the second principle of thermodynamics according to which

Energy tends towards equilibrium of physical entities considered in an isolated system, which is really difficult to accept if psyche is considered.

Or in a more modern sense it can be considered a homeostatic principle through which a partially isolated system regulates itself; in this case a concept theoretically more congruent and not far from the modern complexity theory would be obtained. In any case Freud finds this principle in the secondary process in which energy is blocked, that is to say limited to a certain level.

Constancy is mentioned in *Project for a Scientific Psychology* (1895) as a process which avoids the absolute distension of mental systems, but only in *The Interpretation of Dreams* (1900) does it become a fundamental factor for psychic equilibrium and become identified with the reality principle.

It is still quoted in *Beyond the Pleasure Principle* (1920) but it is defined differently and it seems to be identified with the function which the pleasure principle follows to regulate itself against death drives, constantly busy in destroying and disintegrating organism and psyche.

Freud found himself in the middle of a heated debate while the discussion between Lamarckism and Darwinism was being carried out and this problem has not been completely solved yet.

Lamarck was the greatest theorist of the theory of inheritance of acquired characteristics, according to which there is possibly an orthogenetic force innate in organisms which gives rise to changes connected with the use of organs and which are transmitted to descendants. The biologists of the time like Karl Wilhelm von Nageli and von Baer actually agreed with Lamarck about the existence of an innate natural tendency to perfection. Darwin's contribution ran against these theories, maintaining that chance was the environmental factor which drove evolution. He, however, never stated that chance was the only factor; he thought that it was the most relevant cause of the evolutive processes. Freud always agreed with Darwin about that, the reality principle is nothing but a psychic equivalent of the environmental requests which drive biological evolution.

Rather it was Freud who discovered and insisted on the importance of the earliest environment in the etiology of psychic disorders; according to the seduction theory, pathology is caused by real experiences, then fantasies, and nevertheless genetic traces inherited from the experiences of the previous generations.

In *Totem and Taboo* Freud already proposes a synthesis between Darwin and Lamark and indeed for the first time he outlines the theory of primitive horde and the murder of the father, based on the theory of recapitulation hypothesized by Darwin in The *Descent of Man, and Selection in Relation to Sex* (1871).

This assumption was accepted by Darwin, despite the fact that von Baer's embryological contributions in 1828 based on the study of embryo chick development had contradicted what the first evolutionists such as Friedrick, Meckel, Lorenz, Oken and Augustin Serres had maintained about ontogenetic recapitulation of phylogenesis.

In *Moses and Monotheism* (1938) Freud defends the hypothesis of inheritance of acquired characteristics, as the Oedipus complex has always been considered a structural phylogenetic nucleus which proposes itself ontogenetically in the same form as the supposed primitive experience of the murder of the father.

Commenting on a passage of *The Wolf Man Case Study* he supports Lamarckism and says: "The boy had, here, to adapt to the conditions of a phylogenetic model; and he did, although it did not suit his personal experiences. The threats of castration he had heard were actually expressed mostly by women; the final result however was not delayed too long because of that; nevertheless, his father became finally the person to be afraid of. Heredity prevailed on the accidental circumstances of his life; during prehistory it was no doubt the father who punished , which became later a simple circumcision" (Freud 1914 b).

Later in *Introduction to psychoanalysis* (1916-1917) he reaffirms this idea, connecting it, this time, with the genesis of the primary scene.

Though he showed a certain prudence and caution due to his awareness about the validity of his position in Lamarckian terms, he however exposed himself unequivocally as follows: "I think that these primary fantasies, as I would like to call them, are together with some others, a phylogenetic heritage.

Through them an individual, bypassing their own experience, draws on the prehistorical experience, where their own history is too rudimental. It seems to me totally plausible that all that is told as a fantasy during analysis, seduction of children, sexual excitement caused by witnessing parents engaging in sexual acts, threats of castration, (or better, castration itself) was a reality at the dawn of human history, and that a child, through their fantasy, fills the gaps of their personal truth with prehistorical truth. I suspected again and again that neurosis psychology kept, more than other sources, ancient traces of human evolution" (Freud 1917).

In the supplement of 1918 to the clinical case of the Wolf Man he again promotes phylogenetic hypothesis, balancing Lamarckism and Darwinism through an environmental counterproposal which is not in opposition to the heredity theory: "witnessing parental coitus, acts of seduction during childhood and threats of castration are all surely phylogenetic heritage; these however can be acquired as well through personal experience. In my patient's case, seduction by his older sister was undeniably a real experience; why could not witnessing parental coitus be the same? All that we find in the prehistory of neuroses is that a child catches hold of this phylogenetic experience where his own experience fails him. He fills in the gaps in his construction of individual history with prehistoric elements".

I totally agree with Jung on recognizing this phylogenetic heritage, but I regard it as a methodological error to seize on a phylogenetic explanation before the ontogenetic possibilities have been exhausted and I do not understand why childhood prehistory is refused a meaning which is given on the contrary to ancestral prehistory Moreover I do think it is possible to deny that phylogenetic aims and productions need explanations which personal childhood can give. Finally it does not seem amazing that on equal terms an event which occurred during prehistory and later handed down as an attitude to relive can reappear in the experience of a single organism (Freud 1918).

Here we find, even more evidently, the neo-Lamarckian element of Freud's thought, which distinguished him until his last theoretical contributions and manifested itself in *Totem and Taboo* (1912) encouraged by Darwin's Lamarckian work: *The Variation of Animals and*

Plants under Domestication (1859). On the other hand Freud was always interested in Lamarck, in fact during the First World War, he tried to analyze neo-Lamarckian matters related to psychoanalysis.

The war and the resulting difficulties to get the research material essential to a joint and exhaustive work soon discouraged Freud, who at first tried to hand the work over to Ferenczi.

Lamarck's works examined by Freud, suggest the idea that natural evolution must follow a kind of intentional, volitive, teleological principle, and perhaps this was the aspect of the naturalist's theory which especially interested the father of psychoanalysis.

In particular he finds a correspondence between the omnipotence of thought expressed in hysterical symptoms through drive and the volitive aspect innate in natural processes (Ritvo 1990).

This approach did not contradict even Darwinian Theory; on the contrary it tended to complement it, as the inner constitutive principle (heredity) is coactive with the external constitutive principle (experience).

As is known, the history of ideas in the naturalist field promoted Darwin's theory about random variations followed by natural selection, which again casually rewards the most useful variations to survive in the environment.

Hereditary variations gradually modify the individuals of the species and according to 19th century biology theory of biological heredity these variations tend to be lost rapidly in the species; within a few generations.

Only Gregor Mendel first theorized "heredity units" which mix up from generation to generation, keeping their identity unchanged.

These units were later called genes and their discovery made it possible to hypothesize that random genetic variations actually are not lost within a few generations but on the contrary their characteristics of stability are later chosen by evolution.

Mendel's studies originated therefore that branch of biology called by William Bateson genetics, which on one hand supported Darwinian Theory, on the other took root in the chemical genetics which led to great developments.

Neo-Darwinism originated from the synthesis of Darwinian evolution with the discoveries of molecular genetics.

It still maintains that random mutations of genetic recombination result in changes subjected to a random and gradual selection imposed by environment needs.

Neo-Darwinism is considered by many to be the only evolutionary theory based on scientific foundations and it is a compulsory subject at numerous universities.

According to Lynn Margulis, it is however basically wrong as most neo-evolutionists have a good knowledge of zoology but not of microbiology and of microbial ecology (Margulis 1995). Therefore all information derived from studies of the extinct previous species which highlighted the creative quality of evolution, would be discarded.

The matter of the relationship between evolution by natural selection and that through inheritance of acquired characters has been recently dealt with in very interesting studies.

Doyne Farmer and Rich Bagley have worked on mathematical models of catalytic sets, providing their virtual organisms with different kinds of molecules which can make up the raw material for these sets.

They observed that some organisms could live on nothing but the same molecules previously assimilated. On the contrary others were omnivorous: thanks to a more complex metabolism, that is to say with a higher degree of functionality and organization, they were even able to adapt themselves to new virtual food, keeping their structure unvaried.

Similar natural systems were probably the first adaptive living forms at the beginning of life on Earth.

Autocatalytic cycles are therefore an example of retroactions which assure the stability of an organization at a molecular level. In this way every emerging quality of the combination organism-medium produces evolutive conditions and a context which are the most favorable for the next emergence.

John Caìrns, Julie Ouerbaugh and Stephen Miller altered a gene of E.coli, so that the bacterium would not able to draw energy from lactose. Next they removed the genetically modified bacterium and put it in an environment made exclusively of lactose. The research group discovered that RNA retroacts on DNA in order to obtain a modification for a bacterial offspring able to develop in lactose.

The result is extremely interesting because it reaffirms the idea that an organism is able to actively choose among the mutations required by the environment those to accept and those to reject implying an updated , biochemically founded version of the evolutive theory of inheritance of acquired characters.

Certainly the analogical connection between E. coli and the Oedipus complex is arduous and cannot justify *tout court* any conclusion about the validity of a general evolutive theory compared with another one, but surely it encourages reasoning about the complexity of the evolutive process which cannot be reduced to a simple binary choice among hypotheses.

The serious theoretical limits of Neo-Darwinism lie in its reductionist idea of genome, considered as a linear sequence of independent genes, carriers of stable characteristics of biological phenotypic traits.

Recent research is going beyond these mechanistic limits to propose a systemic concept of genome; that is to say it suggests an approach which tries to understand the effect of the immense complexity of gene interaction, which does not give expected explanations but considers genome a highly creative reality provided with its own evolutive logic.

Evolutive theory is therefore starting to see evolutive change as if it were generated by life's innate tendency to create diversity through sudden changes alternated with long periods of stability, followed in turn by evolutive reconfigurations which are unpredictable and apparently chaotic.

Also "fossil remains clearly show that during evolutive history there were long periods of stability, or lull, with no genetic mutation alternated with sudden and dramatic transitions. Long periods of stability lasting hundred of thousands of years are certainly the norm. The human evolutive adventure started actually with a million-year-stability of the first species of hominids, Australopithecus afarensis.

This new thesis, known as 'the theory of punctuated equilibrium' maintains that sudden transitions were caused by mechanisms which were totally different from the random mutations of Darwin's theory" (Capra 1997).

A new paradigm therefore is asserting itself, as Stuart Kauffman explains: "Much of the order we can see in organisms probably is not the direct result of natural selection but a consequence of the natural order on which selection acted (...) Evolution is not simply striving (...) It is an emergent order respected and refined by selection." Selection can tend towards a certain result but there are limits imposed by self-organization. Therefore selection and dynamic organization work in opposition until reaching a state of equilibrium where selection does not modify anything anymore.

This opposing mechanism is actually even subtler, natural selection in in fact similar to a force causing continuously emergent systems to self-organize towards a non-equilibrium point, called phase transition, in which the system immerses itself again in chaos not in order to lose organization but to permit the emergence of new states which are related to it.

To sum up, a phase transition is that margin of chaos where complex systems amplify themselves devouring disorder and evolving irregularly; it is the ideal place for complex computation.

The margins of chaos or limits of self-organized criticality have been studied in different fields, such as price fluctuations of stocks, irregularities in urban traffic, distribution and frequency of earthquakes (Kauffman 1993).

Molecular biology has tried in the last few years to understand how the non-linear dynamics of the systemic concept of evolution works at its own level of study.

It has been deduced that a random mistake in the self duplication of DNA causing a genetic mutation has a minimal influence and that there are other evolutionary drives which are much more powerful and therefore more responsible for mutations.

Lynn Margulis thinks that there is a symbiogenetic drive which, systemically joining up and coupling microorganisms, causes discontinuous evolutions and mutations with big jumps; the resulting symbiotic identities would be stable in time.

Cooperation and interrelation would therefore be the principal drives for evolution, rather than diverging tendencies as suggested by Darwin's theory.

Nowadays von Nageli and von Baer's hypothesis on the existence of an innate natural tendency towards perfection, supported by biochemical experimentation and a more accurate and detailed analysis of evolutionary history seem to gain credibility. Also in this case it must be admitted that Freud felt that the mechanistic approach was modest regarding the complexity of his object of study and supported therefore both evolutive theories of the time, in the hope of finding a theory which was better suited to the science he was outlining.

In this way he paid the price of giving credibility to the theory of murder of the father and the one of primitive horde which nowadays seems rather naïve; but he never betrayed his deep intuition that the linear reductionist mechanism was unsatisfactory, even though it worked in the case of classical physics and 18th century natural science.

By feeling unsatisfied with epistemological scenarios of his time, Freud anticipated himself; in his own object of study, he saw a new complexity which still raises serious theoretical and methodological questions.

Chapter 4

RELATION VERSUS REIFICATION

> I am not a simple I, an undivided I, an individual I. In me there is a society of individuals who need one another, who divide themselves reciprocally, who are at war and at peace with one another. I cannot ignore the other because "I am" the other, because I am a stranger to myself. I can recognize the stranger as such because I know him in me; I could not preach to him outside me, recognize him outside me.
>
> M. Cacciari

The necessity to treat pathologies different from neurosis has encouraged the development of quite diversified theoretical contributions. This happened some decades after the first Freudian formulations, according to which it was impossible to treat the most serious forms of psychopathology. Therefore various theoretical approaches flourished offering contributions which sometimes were hardly compatible, mostly because no real communication originated among them .Over the years however the need of a synthetic approach has arisen, something that can be assimilated to the proposal of a "comparative psychoanalysis", as Ray Schafer suggests (Schafer 1979).

Even if a real synthetic approach has not developed, nevertheless at least a perspective focusing an increasing interest in personal interaction seemed to have arisen.

This line of development fought its way to the detriment of the classical approach proposed by Freud according to which object relationship was subordinate to drive discharge.

The reductionist temptation still present in the classical point of view gives way to a less simplifying concept of research thanks to its interest in interaction.

In the 1940s drive discharge as a motivating force of behavior was replaced by specific relationship modalities thanks to Harry Stack Sullivan and W.R. Fairbairn.

These two approaches did battle to assert their own ideas; as it was said, there was not a gradual, painless and exhaustive transition towards a point of convergence; it is enough to think of the various theoretical partisan divides originating from that competition, including also contributions by eminent theorists such as Modell (1968) who tends to minimize the points of the approach based on object relations or Guntrip (1971) who, on the other hand , minimizes the points of the drive approach (Model 1968) (Guntrip 1971).

However the object relations theory defines better the nature of the object which Freud already considered the libidinal object and which now differs more and more from the object of academic psychology as it cannot be meant as a discrete and quantifiable entity, but rather a representational anticipation which filters and modifies the past.

The object relations theory reconsiders not only what is objectual , but starts to modify also the idea of what is defined objective, recognizing that it is deeply conditioned by theory, personal interpretation, historical and cultural context. The emerging approach starts to be affected by Kuhn's criticism, which during the 1960s suggested that truth is unknowable and there are only models which are ways of seeing the world, useful but not exact representations of reality (Kuhn 1962).

Psychoanalysis is leaving behind the preconceptions of the paradigm of simplification, and the study of meanings becomes definitely a priority compared to the concern of maintaining continuity with the biological tradition.

The object relations theory does not underestimate the contribution of drive, it rather relativizes it; according to Freud the object is a function of drive, according to this new approach drive is a specific modality of object-seeking.

Consequently interaction becomes more relevant than intrapsychic and with Fairbairn object-seeking is considered a drive which is intrinsically aimed at object-seeking.

We therefore revive two concepts: the concept of object and the one of relation which, from the new point of view, acquire a definitively immaterial and non simplified connotation, with regard to a conceptual abstraction which describes phenomena more accurately.

Freud had already laid the foundation for a primary consideration of relation as object-seeking through his contribution on narcissism.

Indeed "metapsychologically, given the topographic definition of narcissism, the term itself is oxymoronic. Libido is invested in one place or in another; if it is invested in an external object, it cannot be narcissistic libido and vice versa, the concept makes sense only if we assume that once ego is able to capture libido, it can make use of that energy for its own aims, aims imposed on the energy secondarily and by virtue of its associations with the ego.

This argument suggests that along with the introduction of narcissism Freud was moving toward a more active conceptualization of the ego which was not fully announced until the publication of *The Ego and the Id.* The focus on narcissism can be understood as an important early stage in the evolution of the structural model. Even more important, this argument suggests that the concept of narcissism represents another step in the diminishing specificity attributed to drive" (Greemberg, Mitchell 1986).

Group Psychology and the Analysis of the Ego addresses another discussion which gives emphasis to the role played by object relations in psychic structuration, that is to say the relation between the Ego Ideal and the meaningful relation as identification with objects.

It is only with Fairbairn that in the early 1940s a radical change in the metapsychological psychoanalytic structure can be seen: with this author the object relation is primary compared with drive, which acquires a meaning of drive at the service of object relations.

Object-seeking is primary also in evolutive-temporal terms, it is present from birth and auto-erotic pleasure-seeking (narcissistic) is only a consequence of a defective object relation.

The same psychopathology is systematized according to the quality of relations which are regarded deviant and pathologic when they are established with inner objects.

In particular, he says: "as the nature of these relations is the ultimate cause of the symptoms and deviancies of the character, a further aim of psychoanalytic treatment is to effect breaches of the closed system which constitutes the patient's inner world, and thus to make this world accessible to the influence of outer reality." (Fairbairn 1958):

Erogenous zones themselves are vehicles for relations quality. This Copernican revolution therefore aims to build a structure relation model opposed to a structure drive model. Fairbairn ignores the death drive theorized by Freud and distances himself also by Kleinian theory, though he keeps much of its theoretical innovation.

In other words, aggression is the naturally dominant response to frustration which can result in internalization of bad objects (primary identification).

Moreover, according to this author, libido is essentially searching for an object, and this search is so powerful that a child does not give it up even if the relation is frustrating or damaging.

Fairbairn stresses the foremost importance of relation and this concept will be dominant also in ethological and Darwinian approach by John Bowlby.

This author claims that a child's attachment to their mother is not a result of the pleasure obtained through the fulfillment of primary needs but it is primary and phylogenetically determined.

Therefore the sense of trust that a child experiences through the contact with a physical object is primary.

Bowlby's argument, as Greemberg and Mitchell underline, tends to lean towards the practical.

The decisive discriminating factor of maternal actions is her physical absence or presence near the child; this approach disregards the subtleties related to emotive meanings dealt with by the more refined psychoanalytic theories.

However, his attachment theory shows precise references to Fairbairn's idea and to information-processing theory, and, due to its scientific longevity, still now informs modern cognitive theories.

Donald Winnicott, as it is known, focuses on the object-relational abilities, on maternal devotion which can anticipate infant needs and can respond to the environmental stimuli, on her availability to provide a mirror for the development of the personal self.

Thanks to her devotion, the child reaches evolutive goals in the early childhood, thus learns to get in contact with reality acquires a rudimentary sense of the integrated self and of their body.

Winnicott himself describes the features of maternal devotion with regard to functional as a modulator of stimuli.

"In order to preserve the personal own way of life at the very beginning, the individual needs a minimum of environmental impingements producing reaction. All individuals are really trying to find *a* new birth in which the line of their own life will not be disturbed by a quantity of reacting greater than that which can be experienced without a loss of the sense of continuity of personal existence.

It is the mother who lays the basis of the mental health of the individual, the mother who, with her dedication, is able to adapt herself actively to the child. This requires a state of relaxation in the mother which is essential like the comprehension of the way of life of the child which, once again, originates from the mother's ability to identify herself with her child. This relation between mother and child starts before the baby's birth and continues, in some cases, during and after it" (Winnicott 1948 a).

And describing the fortunate meeting between mother and child with regard to the latter's self-sense in relation to reality: "At the beginning the mother lets her child dominate, if she did not do that the breast perceived objectively could not imbricate the subjective object of the child. It is not necessary to say that by adapting herself to the child's drive, the mother gives her child the illusion that what is there is created by them.

These results not only in a physical experience of instinctual satisfaction but also in an emotional union: the baby starts to believe in reality as something of which one can have illusions.

Gradually through the experience of the relationship with his mother the baby uses the perceived detail in the creation of the object he expects. During breastfeeding the mother can repeat this experience thousands of time. She can give the baby the ability to have illusions so well that the child will find no difficulty in his next task, that of a gradual disillusionment" (Winnicott 1948 b).

According to this author a gradual failure of the mirror-role of the mother is healthy, aimed the development of separation and individuation.

The child thus emerges from the condition of omnipotence arisen at "the moment of illusion" when need and maternal relief are consequential and he achieves the ability to perceive realistically himself and others as separate people.

Winnicott believes that the relationship with the environment is so important that he does not hesitate to consider the analyst's emotive receptiveness as a fundamental therapeutic factor in serious psychopathologies.

He says: "There is a vast difference between those patients who have had satisfactory early experiences which can be discovered in the transference, and those whose very early experiences have been so deficient or distorted that the analyst has to be the first in the patient's life to supply certain environmental essentials.

An analyst has to display all the patience and tolerance and reliability of a mother devoted to her child, has to recognize the patient's wishes and needs, has to put aside other interests in order to be available and to be punctual, and objective, and has to seem to want to give what is really only given because of the patient's needs" (Winnicott 1948 c).

In contrast with Fairbairn, Winnicott never explicitly abandoned Freud's drive theory but he considered it secondary to the relational field: in this way he allowed his theory to be widely accepted by the psychoanalytic community.

Moreover it seems evident that Winnicot kept his distance even from Klein's world which had helped him to systematize his research and in this case as well the reason was due to the author's overly emphatic pulsional theoretical bias.

In conclusion, he was always careful to tie down his discoveries to the classical branch, even if in order to pursue this goal, he inevitably distorted the drive theory to allow space for a theory focused on environmental determination.

Kernberg supports a perspective which explicitly tries to integrate these two approaches. He claims to have been influenced by Freud but his perspective draws on Mahler (1980) and Jacobson's (1979) contributions which fully belong to traditional object relations research, even though they are strongly influenced by classical metapsychology.

Mahler speaks about affective-motor libidinisation of the Self which would be a result of tactile and kinesthetic stimulation by adults who, through a positive attitude towards the baby, would help them to build up a basis for the integration of the body-self (Mahler 1975).

Mahler was echoed by Edith Jacobson, who uses the concept of libidinal investment and adds the concept of self as a representation of feelings, behaviors and desires, a factor which is more suitable than ego to be included in a relational approach (Jacobson, 1954).

Through her attempt, she tries to demonstrate that integration is not only possible but also necessary in order to avoid hesitating in divisive and partial positions; nevertheless she is not free of contradictions even in her most recent formulations and hypotheses.

According to Kernberg, drive provides internalized relations with an emotional component and for this reason he does not distance himself from the classical approach, even if he does not exclude the reactive nature of aggressiveness resulting from frustration.

He seems rather contradictory when he claims that affections, even anger and sexual excitement, are highly complex realities rather than simple discharge processes.

Their complexity is due to the fact that they are a result of the activation of an inner object relation and always have a cognitive component, even when they are chaotic affections of visibly regressed patients.

Internalization of inner primitive object relations is facilitated by affective experiences which Kernberg defines “apical”, they occur in highly pleasant or painful situations and require a high level of arousal; in this way affective structures of memory take shape, therefore the tripartite structure of Id, Ego and Super-ego. Internalized object relation would originate from the first internalized object relations.

Affective subjectivity itself would form thanks to the first experiences of pleasure and pain which, once they have been activated through perceptive associative links and conditioned reflexes, would be a precursor of the earliest symbolic thought.

Kernberg makes a wide use of neurobiological research to support his own theories on the development of thought, and also to explain the origin of the dynamic unconscious and of the first fantasies of frustrating or bad objects which include attempts to expel unpleasant sensory experiences, resulting in the so-called projective identification. On the contrary pleasant experiences would be introjected through fantasies of incorporation which are also deeply rooted in the intersubjectivity of the relation with the real maternal environment. However the apical affective states organize inner object relations not only through positive feelings but also through unpleasant and frustrating experiences.

Fraiberg and Galenson's studies examine precisely the matter of bad object internalization in agreement with Kernberg's contribution (Galenson 1986).

The excessive aggressiveness which may result interferes with the integration of dissociated relationships divided into completely good or bad, and may cause problems in the development of object constancy (Kernberg 2000). These considerations show that Kernberg is intent on connecting Klein's contributions o an observation on objective bases linked to the quality of external contextual relationships ; moreover the author seems focused on defining his concepts with such precision as to guarantee continuity with the classical theory.

Theorists belonging to the so called “mixed model” have pursued integration not only in theoretical but also in clinical experience.

Heinz Kohut is surely one of the most influential; he solves the question of the interpretation of drive, ascribing to the Self the functions of energy regulation which the classical model attributes to the Ego, Id and Super-Ego.

Hartmann had already spoken about the Self as a representation produced by the Ego similarly to the presentation of objects (Hartmann 1950). Upholding the hypothesis that investment of libido concerned the Self instead of the Ego, he tried to update Freud’s structural model in order to stop the spread of models which would result in the object relations theory (Wolf 1988).

Typical terms of the drive model appear again: for instance Kohut, regarding *narcissistic self,* claims that “its narcissistic cathexis (...) is retained within the nexus of the Self and does not make that specific partial step toward object love which results in idealization”.

The Ego ideal is prevalently related to drive control while the narcissistic Self is closely interwoven with the drives and their inexorable tensions, therefore Kohut concludes that: "the ego experiences the influence of the ego ideal as coming from above and that of the narcissistic self as coming from below".

The author insists on combining the concept of drive with that of the Self, of narcissism and of the psychic needs of the structural model, maintaining that "while narcissistic urges may be considered as the predominant driving aspect of the narcissistic self, the grandiose fantasy is its ideational content (...) and that in the maintenance of the homeostatic narcissistic equilibrium of the personality , the interplay of the narcissistic self, the ego, and the super-ego may be depicted in this way.

The narcissistic self supplies small amounts of narcissistic-exhibitionist libido, which are transformed into subliminal signals of narcissistic imbalance (subliminal shame signals), as the ego tries to reach its goals to emulate external examples and to obey external demands, or to live up to the standards and especially to the ideals of the super-ego"

The author even relates conceptually narcissistic libido to the problem of the relationship between Self and psychosis: "The ability to accomplish an act of brutal mutilation depends in some cases on the fact that the organ which the psychotic cuts off has lost its narcissistic libido investment, that is to say it is no longer a part of the Self and therefore can be removed as if it were a foreign body" (Kohut 1978)

In this way the Self becomes a real psychic entity and not merely a mental representation.

The Self is therefore a place for structuring relationships and affection laden drives; the inevitable failures in care would be modulators of a healthy and structuring development.

In particular Kohut claims that "the new-born infant who is likely to survive psychologically is born in a human environment of empathic-responsive self-objects, the same way as, in order to survive physically he must be born in an atmosphere containing a normal amount of oxygen. His budding self is pre-adapted for an empathic environment responding his wishes and needs like his respiratory system is pre-adapted for the oxygen contained in the surrounding atmosphere" (Kohut 1977).

However, Kohut's construct is not without uncertainties and rather acrobatic attempts to adapt terms and concepts to the drive model.

Note for example the implicit simplification in this passage: "the most important source of a well-functioning structure is the parents' personality, more precisely their ability to respond to their baby's drive needs with a non hostile firmness and non seductive affection" (Kohut 1978).

Is it really possible to reduce the complexity of narcissistic mirroring which structures the self to an adaptive response to drive needs?

The question then arises as to what drive actually means-Are drives then also the needs of self-objects, such as the need to feel successful, recognized, admired, respected, the need to experience a benign opposition, to experience a similarity with the self-object, to merge with the idealized object, to be able to provoke responses from the object to one's needs. *Also the question of aggressiveness as a drive is not less uncertain as an adaptation to the classical model when self psychology considers it "a product of disintegration of an environment which is indifferent to birth. This concept assumes that the angry child has, from their point of view, a reason to be this way and that their *anger is not the expression of an innate aggressive drive toward gratification.*

After discovering that aggressiveness can dissipate through an empathetic approach to how the environment (including the therapist) turned out to be insensitive, possibilities to abolish destructive aggressiveness in individuals and in the world begin to take shape.

> Hostilities between nations, for example can be considered not as the predictable discharge of an innate drive which condemns mankind to a final nuclear extinction, but as comprehensible reactions to an indifferent world, for example to an environment threatening total destruction"
>
> (White, Weiner 1986) (my emphasis).

In 1971, for example, he distinguishes narcissistic libido from object libido; the former invests self-objects, in which the object is perceived as an extension of the self, the latter would invest external, real and separate objects; The disorders of the self would be expressions of problems related to narcissistic libido, while object libido would remain in the area where impulses act, that is to say, wishes derived from drives.

It follows that, from the perspective of a sequential development of psychosexuality, the achievement of a cohesive self is a condition for the development of an oedipal conflict.

Conceptual models are thus related to two different periods of development, maintaining both approaches.

However, considering the complexity of real phenomenology and Kohut's theoretical system, this distinction seems artificial and forcedly aimed to save both approaches without displeasing anyone.

In *The restoration of the Self,* Kohut considers drives also a product of disturbed primary relationships; how can this be reconciled with the assumption that structural neuroses do not imply any pathologies of the self? Kohut tries to resolve the contradictions of his theory proposing two pathologies of the self, an oedipal pathology and a pre-oedipal one, but doing so he offers an alternative to the classical theory, which explains the oedipal matter as a drive conflict, and corrupts his project of integration. However, it must be admitted, that Kohut's contribution, in agreement with Lachmann and Stolorow's theories, underlines how the interruption of the development of defense is a decisive factor for serious psychopathology, in contrast with the models based on intrapsychic conflict (Stolorow Lachmann 1980).

The question of structural defect is therefore posed again by the self- psychology even with the inevitable contradictions that integration between theory and clinic may imply.

Kohut tried to remain faithful to the idea of a neutral analyst until his last contributions, and then he reluctantly admitted the unavoidable influence of the observer on the observed (Kohut 1982).

By proposing empathetic mirroring as a specific therapeutic tool, self psychology widens the range of analytical techniques in the treatment of serious psychopathologies and also extends the field of observation of analytical interaction to the impressions which the patient develops of the analyst and the analyst's behavior towards them.

The positive effect of a loving atmosphere on the child's development had already been expressly anticipated by many theorists , as White and Weiner remind us, and by many analysts such us Anna Freud, Heinz Hartmann, Erik Erikson, Margaret Malher, Edith Jacobson, Renè Spitz, Anna Ornstein, Alice Miller, Louis Sander, Marion Tolpin, John Mack, Joseph Lichtenberg and T. Barry Brazelton.

The concept of the psychotherapeutic atmosphere as self isolating and as a facilitator of the maturing processes was also at least implicitly suggested by Carl Rogers's humanistic psychology long before Kohut's discoveries.

We actually owe to Kohut a systematic understanding of how empathy decisively contributes to the development of the self both in the field of primary relationships and in the one of therapeutic relationship.

In this way the author speaks about a baby who is really different from the one Freud studied, not only for metapsychological reasons but also due to environmental factors, determined by a family context, significantly changed from the one which Freud studied at his time.

Freud's patient lived in a highly stimulating context, where affectivity was mixed with eroticism, also because of exaggerated proxemics; Kohut's patient is an ex-baby often ignored by adults, busy in a frantic world which leaves little space for emotive involvement and for mirroring (White Weimer 1986)

Like Kernberg, Kohut, too, leaves many theoretical questions unresolved, which make a convincing integration of the theoretical approaches difficult.

In order to overcome the difficulties caused by these attempts, Joseph Sandler adopts a different method, which "grafts" the theory of object relations with the classical drive theory.

The goal of drive does not exist in a vacuum, in the abstract. It arises as a wish, and a wish always implies fantasies about the self and fantasies about the object.

The relationship gratifies the drive through the discharge of physical tension but it also fulfils the wish as a specific relational need.

It seems a good attempt, even if Sandler over time modifies his position to leave more space to the relational perspective; drive based wishes are indeed considered, from a certain point on, as a category of desires which are, generally speaking, mainly dependent on the quality of the relations.

Indeed, Sandler speaks specifically of desires for interaction, explaining in this way the primary relational nature of desire and its independence from drive (Greemberg, Mitchell 1983).

In conclusion also Sandler has difficulty finding a unifying theory, but this does not mean it is not possible to find one. In the last few years Daniel Stern's studies have greatly contributed to the validity of object relations theory; he has deeply examined the mother-child interaction from the very first moments of relational life. According to this approach, mother and child form a real object relation (between differentiated objects) in the very early stages of development (two-three months) not only through the mouth, as is emphasised by the drive theory, but also via the hands and the oculomotor system, by which the baby is able to manifest a proto-intentionality keeping or averting their gaze from the mother. The mother-child interaction is almost always ruled by both of them, who thus precociously start a battle for dependence and independence long before the exchanges of the anal stage. A very active child emerges who "is able to adaptively interact with the environment", who is never autistic, as he has always been engaged in social relations and been interested in a visual, aural, manipulative incorporation of the surrounding reality (Rossi 1988).

According to Stern the active function of memorization is already present in a child even before birth; therefore there would be a remarkable continuity between prenatal and postnatal experiences as if birth did not affect the continuum of mental growth.

Moreover affective and cognitive functions already well differentiated at the age of three months would be enabled by "amodal perception", consisting in translating an experience from a certain sensorial modality into another modality, allowing integration of different sensorial experiences and their derived representations (Stern 1987).

Transmodal functioning allows the performance of transmodal transformations, through which an experience from one sensorial modality could be recognized in another thus constituting a sensorial unity resulting in the identity constancy of an experience in time and space.

It is precisely this identity constancy which unifies our perception of the world, allowing affective intersubjectivity and an early form of pre-verbal awareness (Mancia 1988).

The validity of neuropsychological and observation research into object relations theory consists precisely in confirming the existence of the intersubjective relationship even at the archaic levels of personal evolution.

Moreover Stern's observations highlight some new features of the mother-child relation indicating the reciprocally processual aspect of experience structuring.

This processual function goes beyond the concept of restraint or reverie, still linked to the objectifying, reductionist and intrapsychic drive theory. Reciprocal regulation or attunement does not describe a simple transformation of the stimulus-response type but a specifically bidirectional adaptation. Initially oriented towards the mother's affective mirroring of the child, it is contingently followed by a maternal affective-emotive variation, through which her expectations manifest themselves. The message returned by the mother is inserted in a self-regulating dialogue which creates new experiential bonds in each protagonist of the interactional scene (Parisi 1989).

Through these considerations we arrive at a scenario conceived as active-evolutive and self-organizing of the relationship which structures the mind (the minds).

The relationship thus conceived can disregard the explicative ontological level to arrive at a more effectively informational and descriptive heuristic model.

Emde however reminds us that Freud already anticipated the regulative (informational) function of affection. Distancing himself from Helmholtz's reductionist-mechanistic model, in *Introductory Lessons* (1916-17) and completely in *Inhibitions Symptoms and Anxiety,* affections are real self- regulation signals of the working of the mind (Edme 1980).

It goes without saying that in Freud this concept anticipates Stern's interactional and informational theory. Stern's contribution certainly places itself as an outpost for an interdisciplinary study of the mind. In fact interactional and informational vertices are shared by both Piaget's evolutive psychology and the modern psycho physiological models.

They seem to shape "an epistemology which is correctly built as it at least requires the hierarchization of the abstraction levels of the used conceptual apparatuses, permits verification of the validity and consistency of the superordinate general theory and promotes the development of a metatheory that (...) *allows* the regulation and the control of the general theory via an appropriate metalanguage (...) Moreover, it is important to note that , *as with* Freudian system, the correspondence between the abstract and concrete levels of experience, between practice and theory is so high that it permits the theoretical activity to directly constitute an optimized system of evaluation aimed at verifying the awareness level concerning the falseness and truth of experience, and at the same time to constitute an active process which enables a learning function of the experience itself" (italics mine) (Corrao 1989).

Generally speaking, the structural model of relationships seems to be more complex than the drive model, it follows an inclusive logic, hence an hypothesis does not necessarily exclude others, there can co-exist different points of view , therefore, in principle, incompatibility between drive and relation can be absent, or between a theory of temperamental differences and a theory of relations.

It is consequently possible to interpret the normality and the pathology of psychological behavior from the superordinate point of view of a theory of subjectivity and its vicissitudes concerning the relationship with objects; this is indeed not incompatible with a metapsychological vision which considers the subject a psychic apparatus busy with objects aimed at drive gratification (Ferreri 1992).

A new idea can start making way: the preference for one or the other theory simply denotes the choice of an exclusive point of view of phenomena, which can simplifyingly favor a causal category, leaving any other implication by the wayside.

The working of the mind can be conceived in terms of energy or of information, in terms of instinct or in interpersonal terms or through any other useful metaphor; it is only an epistemological and methodological matter (Siciliani1970).

Our choice can become exclusive if we become aware of handling metaphors which, however sophisticated and fit for the purpose they may be, still remain only representation, categorization or modeling systems none necessarily in contradiction; as metaphors and models they are containers of concepts whose exhaustive definition is actually impossible.

Yet there is someone, like Thorndike, who believed he could provide a complete model of human instincts making a list of forty, or like Bernard, who arrived at five hundred, without completing it; all this despite Freud's theory according to which even if there are certainly two categories of instincts, that of life and that of death, they are to be conceived as groups of many others.

Even though Freud speaks of drives and not precisely of instincts, the common energetic metaphor and the same destiny of complexification and abstraction permit considering these two concepts conceptually analogous.

More precisely instinct models should be considered as including different, complementary and synergistic instinct levels, which are more or less complex, that is to say more or less informed with noetic meanings.

Monakow, for example, already many years ago, suggested a distinction between primitive forms of instinct and forms which had been processed by the noetic sphere, deducing a condition of the mind multi-motivated and organically imbricated.

In other words Freud's principle of over-determination is valid here, it follows that it is difficult to consider that a behavior is determined by a single instinct; instincts integrate each other, come undone and present themselves in different functional, in any case, extremely articulated aggregates.

The same can be said of the contrast between the instinct model and the relational model, where the dualism is only apparent, and it is due to the fact that there are particular aspects of psychic life which use models more suitable to their specificity.

The necessity to use two cognitive perspectives, typical of two different autonomous disciplines, in order to completely grasp a phenomenon is not an epistemological scandal, especially for those used to reasoning in the psychiatric field where, as Guiraud would say "there is a double-faceted monism", which in respect of the different disciplinary categories, allows a better knowledge of causes (Balestrieri 1998, Guiraud 1950).

To sum up, these theorists met a difficulty which is more epistemological than metapsychological; it may be necessary to keep a perspective based on an inclusive logic, instead of an exclusive logic which searches the truth only through one or another theoretical approach.

It is true that science must be essentially aimed at identifying distinctions otherwise no scientific reasoning would be possible, but it nevertheless should take into account also what draws together, what integrates, not in order to identify first or unifying principles, but to reveal equivocal models of interpretation of the phenomenal reality.

It is no longer a question of making different perspectives homogeneous and consistent, but of understanding how distinct points of view are reciprocally compatible and complementary.

From every point of view one can observe that which occurs in various outcomes as generative of other vertexes "there is no world but which we experience through those processes we are given and make us the way we are, We are in a cognitive domain, and we cannot either leap out of it nor choose its origins and modalities. (...) We cannot univocally connect a determined experience to its origins. Actually, every time we try to find out the source of a perception or an idea we find ourselves in a fractal which continuously repeats itself and every time we try to analyze it, we discover it is full of details and interrelations. It is always the perception of a perception of a perception....the description of a description of a description... (Varela 1984).

Drive and relationship, relationship and drive determine each other as vertexes which the observer chooses from time to time according to their interest in each specific case. They do not fight but reciprocally configure each other; there is no monopoly or hierarchy of any vertex but a democracy where the only cognitive possibility consists in accepting the trauma of "trans-inspection", that is to say the willingness to accept other vertexes as sometimes being more suitable to the phenomenological typology under consideration (Maruyama 1976).

The borders between objects of analysis and system hierarchies exist only in the observer's mind and are the result of their decisions and their more or less conscious choices.

The completeness of knowledge is an asymptote, a vanishing point for knowledge under the pressure of an epistemic drive. But the studied object is complex, hence the opposite of complete.

Many critics think that complexity substitutes the desire to know everything, the result of a compulsive drive possibly of neurotic origin.

Instead, the opposite is true: complexity is the awareness that nothing can be included in an exhaustive definition, as the possible observation vertexes of a single phenomenon are potentially infinite. There are no *pontifical* metapsychological models "privileges, thrones, epistemological sovereignties" as Morin would say, with figurative verve but glimpses of unknown lands which can be explored with maps and strategies requiring continuous reinvention.

Thus is sketched a transverse, trans-disciplinary, interactive thinking, which tends to close wider and wider explicative circles and whose last aim is to definitively close thought in an omni-explicative super circle .

Wandering thoughts endlessly tangle and loosen and the last word is never pronounced.

Chapter 5

CAUSALITY

I would like to resume on the matter of causality. It is a very important issue either from a philosophical or psychological point of view; indeed, it is one of the concepts which require a multi-disciplinary analysis to be grasped, as Piaget's genetic epistemology and the following studies show.

For this reason I will say just a few words which can be useful to our aims, suggesting that a deeper analysis, extremely interesting also for our discipline, could be undertaken by those it may interest.

The first great current of thought treating causality dates back to Aristotle, according to whom every change in the natural world is originated by four causes: material, efficient, formal and final.

The formal and the efficient causes have crucially characterized western scientific and philosophical thought.

The formal cause concerns the form of scientific law which rules the logical coherence between events, and the manner in which it is conceived in the mind of the scientist.

The efficient cause consists in the concept of ultimate cause of an operation, in the usual sense of the word, as it is contained, for example, in the explanation of the operation even of a very simple mechanism. A hammer blow causes (efficient cause) the penetration of the nail into the wood.

When, during the XVII and XVIII centuries, the field of mechanics, based on the idea of elementary particles of matter prevailed, scientific thought developed the conviction that it was possible to explain natural phenomena through an efficient cause which was traceable in matter just as for any man-made machine.

With Newton the importance of mechanism in the strict sense tends to be relativized because he introduces concepts such as gravitational force, which are difficult *to materialize* that is to say to be explained as they are properties directly connected with particles (Khun 1971).

With gravitation there must be a distant cause, not mediated but any material means, therefore how can one explain the concept of efficient cause, of *final material cause?*

Dealing with the phenomena of magnetism, electricity, and heat, the problem arises even more seriously: how can one determine the cause? The cause of the penetration of a nail is a hammer blow but is the cause always the hammer blow or does it relates back to something previous?

And what if the hammer does not possess material characteristics?

This was one of the theoretical problems Freud had to face early on: explaining behavioral anomalies, that is, psychopathologic symptoms through the concept of efficient cause (isolated or first cause).

If it was difficult, as he soon nostalgically realized, to refer to material particles of the neurological substratum, it was even more improbable to take an impalpable concept such as psychic energy, which however, as we know, Freud abundantly did.

However it was necessary to start somewhere, to search for evidence. This necessity was becoming urgent as electricity, magnetism, optics, and thermodynamics, were provided at least with a mathematical form, from which it was possible to extrapolate a formal cause with a certain elegance and a verifiable predictive faculty which was difficult in the case of psychic energy.

Recently Halbwachs has insisted on reminding us that there is an ontic level, that is to say a world of real, physical, objects determined in time and space (Zingarelli 2001). It is distinguished in specific domains such as "the scale of certain magnitudes related to matter, like mass, energy, velocity and the orders of magnitude typical of each level, are so different from one level to another that it is possible to study each of them independently of the others with an excellent approximation, even though, of course the distinction is never absolute.

This seems to be a specific trait of the *physical* aspect of the world which results in the formation of different, relatively independent, physical sciences .This trait cannot be found in biological, psychological, sociological, etc. ...sciences" (Halbwachs 1971).

Early on Freud, perhaps, stumbled on explaining the concept of causality, erring about the genesis of neurotic symptomatology through an efficient cause (ontic level), confusing the epistemic level with the psychological one. Also because when the subject studies themselves, as happens in psychoanalysis, they interpret the connections between the phenomena they observes as a priori structures which the mind imposes on sensations (Kant) or interpret the representation of their own object (psychic workings) following "the pattern of voluntary action revealed by introspection, and in this case will provides the prototype of the cause (Maine de Biran)" (Halbwachs 1971).

Piaget has extensively demonstrated (...) that the more o less anthropomorphic, more or less formalized impression of a *capacity,* always related to the causation and which alone gives it explanatory power, its psychological significance, precisely originated from this primordial pattern of one's action"; that is to say it is already present in the child's mind and represents a natural tendency to homogenize reality, to assimilate it to a known idea in order to make it more intelligible.

Traces of this need of intelligibility remain nearly unvaried in the adult researcher's mind and the anxiety to know subsides only by creating models like that of causality: intelligible, aesthetically perfect and also for this reason scientifically acceptable by the community of researchers (adults, all former children).

It is in the context of our models and inasmuch as they pretend to go beyond the simple experimental recording (of *sequence of phenomena)* (my emphasis), that is to say in the context of every particular physical theory that we are led to explicitly introduce the distinction between events-causes and events-results explaining in the expression itself of physical laws that the former produce the latter. The question is not easy to resolve even in physical sciences, as causality indicated through the relation C →E (C causes E) is nontrivial. Other causes or conditions may contribute to generate E., as Burge explains, "one should discard the case in which factors different from C are necessarily connected with C, for example when rubbed on a rough surface (cause) a match will light (effect) only if it is dry (first condition) and if the event takes place in air or in oxygen (second condition). As the conditions are as necessary as the cause, one cannot speak of a single cause but of a causal complex system" (Bunge 1959)

In fact, causality in nature consists always and only of "a partial and unilateral representation of reality and acquires explicative value only as a first and elementary approach to phenomena and which must be completed and *combined* with other approaches if one wants to formulate a complete and consistent theory of the studied phenomenon when it arises at an ontic level".

Back to Freud, much is fairly forgiven the pioneer, as both the language and the representation of reality are not flexible tools, especially in certain areas, in particular we cannot forget that causality is an abstraction which disregards a range of intervening factors; moreover Freud actually arrived quite soon at the concept of multiple and circular causality. The development of the concept of causality is to be inserted in a wide range of theories developed by the child on the workings of reality and on those practical, concrete and intuitive rules, which help him to predict his own and other people's behavior. They are often wrong, above all if they are related to the most archaic levels of development.

At a primitive level, the child between two and five years of age begins to distinguish between desire and beliefs; reality begins to assume its own operation which frustrates the child's magical expectations but also forms the basis of more elaborate and complex theories.

These appear at school age, during pre-adolescence and on the basis of these new conceptualizations, reality can be interpreted in different ways, it is possible to think from more than one point of view.

With adolescence the question becomes more complicated; different opinions are formed and compared with the objectivity of the facts, therefore only a few points of view can be considered true (Chandler 1988; Flavell 1988).

It is evident that these theories of the mind condition the perception of reality and this, in turn, is conditioned by primary attachment to the mother, who with her emotional drive makes one or another of the theories more meaningful For example a mother's intrusive behavior can contribute to the theory that thought can be read by an external observer or a child whose mother does not respect the difference between her wishes and her child's could encourage the univocality of a strict unidirectional and categorical description of the opinions (Spence 1993).

Consider how a particular development can affect a theory of causality and rationality, if they, too, are influenced by the communicative-emotive quality of primary relationships.

Rationality and causality seem also to be related to the reliability and predictability of events which follow actions or modalities of reaction; just as repeated unreliability related to meaningful experiences is itself an organized experience.

The child is a victim of circumstance only in the sense that understanding what they are undergoing is not within their capacity but they feel it and adapt to their environment, searching, however, for that experience of efficacy consisting in self-regulation through subjective anticipation of the fact that the environment will not participate in the regulative process. (...) For example a child who has already undergone notable self regulatory inclinations, when they see that their mother is nervous because they have the flu, may relate the first experience to the second more or less intensely, in the unconscious belief that their demands damage their mother (Seganti 1991).

Therefore the child may develop a theory according to which they are the cause of parental uneasiness, tracing the apparently rational cause to them.

Freud had already amply discussed this in Civilization and its Discontents, where he arrives at the conclusion that the sense of guilt is a superego reaction to the aggressiveness repressed in the ego.

The sense of guilt turns from collective necessity into individual unease. "Therefore it can be deduced that also the sense of guilt produced by civilization is not recognized as such, it remains for the most part unconscious or arises as unease, a discontent for which different causes are traced.(...) Sense of guilt (the harshness of the Super-Ego), therefore, is the same as strictness of conscience, it is the perception the Ego has of being thus watched, the reckoning of the tension between the desires of the Ego and the needs of the Super-Ego. The underlying anxiety, that is to say the fear in the face of this critical component, the need for punishment, is a drive manifestation of the Ego (which has become masochist under the influence of the sadistic Super-Ego). That is to say it is a part of the drive present in the Ego, striving for inner destruction, invested to create an erotic bond with the Super-Ego".

I think that these archaic theories to a certain extent also affect the theoretical position of adult minds, even those of researchers, psychotherapists and analysts'.

They are to be recovered and analyzed by the analyst himself in order to be more objective in the exploration of other people's minds. I thus reverse Fonagy's perspective which claims "the patient's capacity to recover a theory of the mind is therefore the first and most important act of the analytical process" (Fonagy 1991). As far as causality is concerned we can conclude that "this abstraction is (often) (my italics) correct, at least as a stage of the analysis of real relationships and, indeed, as we explained, the pattern of the simple causal relation is the most evident basic form of explanation, the most satisfactory, at least in the heterogeneous form. But we must go further, as we must explain not only simplified abstract models, but also reality in its entirety.

More precisely, the isolation and recognition in the apparent chaos of innumerable real relationships, of a simple causative relationship, offer a partial explanation of this reality; they suggest a criterion for putting chaos in order" (Halbwachs 1971).

Chapter 6

LOOPS

The hope of Freudians that the psychic field will one day be included in the neurological domain is a false problem which increasingly restricts and embroils itself; hence we are not as certain as Hartmann is that "the fact that psychoanalysis has a biological basis is its most solid methodological advantage" (Hartmann 1973).

Actually, what is psychic identifies a biological aspect, while the latter identifies a psychic aspect resulting from its functional organization.

It is therefore necessary to completely accept what doing without Freud's paradigm of simplification implies, by integrating linear logic which recognizes that the psychic domain is autonomous from that biological, with a circular logic which considers it connected but not confusable with the biological domain.

In other words, the psychic is biological and the biological is psychic, but the two domains are not reciprocally reducible.

From a new logical vertex this is not a language artifact, but an antinomy, a paradox; nevertheless it is the only possible nature (reciprocally justificative between two levels of phenomenal reality) of a heterogeneous and complex object. We come across such an antinomy every time we study the relations between the different knowledge domains; it reveals *a fortiori* the very nature of objectivity, built on an endless dependence.

Therefore, scientific truth thus seems more similar to a loop than a part of a line with a beginning and an end, and the hypothesis of an essential relationality of knowledge, which cannot justify itself through apostasy, becomes more credible. More generally speaking "our perspective takes into consideration a sort of deutero-learning which can make relationships possible and bearable.

In this perspective the use of certain languages, ossified, hydraulic, mechanistic, energetic, stiffening, technical, initiatory, self- referential, computer... languages more frequently adopted in the past is of no use. Languages which describe the living in metaphor, through a classification of feeling, being, understanding of the human regardless of its specific characteristics related to pain, need, hope and affective-cognitive signification" (Ceruti M., Lo Verso G. 1998).

Truth has to be conceived as empirical-relational without ontological problems, nourished with the relativities of the real.

That's what Edgar Morin decisively states: "The relativity of the reality implies not dissociating reality from unreality any more, not searching for an ontological fundament anymore, but conceiving reality as a relational datum, which clearly originated at the beginning, from the relationship between the man and the world, and above all, conceiving it not as a simple but a complex datum" (Morin 1969).

"On the whole, scientific undertaking could be much closer to the variety and richness of arts than logicians have supposed" (op. cit. c).

In agreement with this perspective, I object to the approaches such as Willy Hockeppel's treating the questions of scientific truth as if they were abstract formulations, cut off from any tradition and historical context.

These realities not only affect the research developments but also inform them to such an extent that they change the truth over the historical periods, "it is an immanent element of history, not an objective external goal (op. cit. d).

The fact that in a certain historical period a point of view on nature is prevalent should not lead to believe that we discovered reality.

It only means that nowadays other forms of reality have no customers, friends or defenders yet and not because they have nothing to offer but because they are not known or their products are not appealing (...) in other words there are not only art forms but also thought forms , truth forms, forms rationality and therefore forms of reality.

Anywhere we turn we find no Archimedean point of support, only other styles, other traditions, other order principles" (op. cit. e).

The hidden side of the loop is reassessed; quality more or less prevails on quantity, according to the chosen artistic style and demands more dignity, requiring research to reject to the arrogant omnipotence of numbers.

Kierkegaard had already reflected on the fact that thought was indissolubly connected with the thinker's subjectivity, that is to say with its potential creator.

Creativity and representation of reality are not antithetical; they are the two sides of the knowledge loop.

Feyerabend's epistemological anarchism is therefore non-conformism which liberates, going beyond the limit imposed by method, conceiving different possible methods; "while a political or religious anarchist wants to abolish a certain life form, an epistemological anarchist may want to defend it as he has no feeling of eternal loyalty or aversion to any institution or ideology (Feyerabend 1975). Speculative forays into places on the edges of certainty challenge knowledge to continuously redefine itself and to proceed in order to expand the thinkable space; we undoubtedly need a dream world so that we can discover the features of the real world where we think to live in".

Also for Kurt Huebuer, professor at the University of Kiel, "the source of scientific progress lies neither in abstract rules of falsification nor in inductive or similar inferences, but in the whole mental and historical situation in which a scientist finds themselves (...) It (the philosophy of science) tries to solve its basic problems-the nature of the methods to apply and the theoretical justification of the propositions obtained - through simple reflection , where thought is abandoned to its resources and to sophistication..." (Huebuer 1969).

Once again, as Gottfried Benn remind us, "even though science is on the whole damaging it is always possible to learn through it", on condition that the strictest rules which regulate research do not smother the creative impulse of scientific thought. Thinking in complexity terms, implies also that, it does mean declining a new paradigm, but applying a thought style which rehabilitates reasoning about knowledge nature, which is ,after all , the only reasoning we can consider scientific (Morlotti , personal communication).

To sum up, "science, as it is, is one of the numerous thought forms developed by man and not necessarily the best one. It is loud, noisy, impudent, and it is superior only for those who have already decided on its favor, or have accepted it without examining its advantages or its limits (...). Thus science is much closer to myth than a scientific philosophy is prepared to admit" (Feyerabend 1975).

Going beyond Popper's approach excludes the idea of an objective knowledge, indeed psychoanalysis is not based on falsifiable concepts.

This does not justify thinking that psychoanalysis is a construction, a play which does not reveal anything of the real, but creates a particular real different from empirical reality.

It is not comparable with mathematics as it is conceived by Wittgenstein but its object is closer to Einstein's concept of mathematics as a conceptual tool which is suitable to describe the empirical, or to Godel's assumption according to which there always will be classes of mathematical statements which are true but unprovable.

Godel's theorem "introduces a discrepancy between the sets of true mathematical statements and the sets of provable mathematical statements : all provable statements are true , but not all true statements are provable, as however much the original set of axioms and of inference rules is enriched, there will still be true but unprovable statements. Theorem proving strategies are always finite while mathematical truth is infinite" (Benvenuto 2001).

It is true that part of current psychoanalysis, the constructivist approach, seems to wash its hands with regard to the value of the truth of its work. As nothing real comes into play, only the consistency of the play matters, but in these terms the argument does not hold together, it would be like asking a patient to invent inner scenarios for pure *entertainment.*

Empirical reality indeed exists as distinct from objective as well as mathematical reality and this direct and analytical experience is a reality which can be subjectively experienced but is neither quantifiable nor accessible and it is only indirectly accessible to the observer's senses.

It is inevitable that the psychic reality is grasped through narrative constructs which, however, do not deny the real consistency of what they describe.

Algebraic formulations to which authors like Bion or Lacan turned have a different meaning: they conceive mathematics as a metaphor which grasps the unspeakableness of the subjective, abstracting a synthetic, operational figure, even if it is not a matter of quantified entities.

It is therefore "representative of qualitative relations and cause-effect connections between hypothesized mental properties and behavior."

The cause-effect connection is here to be understood in a broad sense as we do not have mechanical elements which cause simple deterministic effects; on the contrary, as we will see, the relation has many causes or is overdetermined, as Freud himself had well understood. One of the acceptations of the term overdetermination in Freud refers to the fact that a single formation of the unconscious relates back to several determinants. Although this sense is the least used and shared, it is certainly the one which gets closest to the meaning of multicausality and by analogy to the meaning of probability used by Boltzmann and in quantum mechanics.

The matter also concerns overinterpretation, according to which a single oneiric image, for instance, is the result of the condensation of a series of unconscious thoughts, hence interpretation has to touch overlapping layers of significance.

Freud can therefore add further interpretations to one which already has "its own consistency and value". It is possible to think that a part of the non-interpretability of a certain formation is implicit in the same concept of overinterpretation, as Freud reminds us speaking about the dream's navel, in a well-known passage from The Interpretation of Dreams. reud wrote, as Laplanche and Pontalis remind us "There is a passage in even the most thoroughly interpreted dream which often must be left obscure because we realize that in the course of interpretation there is at that point a tangle of dream-thoughts which cannot be unraveled and which moreover adds nothing to our knowledge of the content of the dream. This is the dream's navel, the spot where it reaches down into the unknown. The oneiric thoughts which we encounter in interpretation cannot, from the nature of things, have any definite endings; they are bound to branch out in every direction into the intricate network of our world of thought" (Freud 1899).

Chapter 7

DATA COMPLEXES

A data complex is a complex datum that is to say the result of an infinite intertwining of assumptions, preconceptions, and cultural determinants.

Data is complex despite rules leading the research, but the rules are quite often reformulated by new procedures and new intuitions. It is presumable that even political pressure could modify a data complex in favor of one or another theory.

Plato was first in critically dealing with this matter; we only need to think of his objections to *antilogike* in *Republic,* 453 and in *Theaetetus* 164c.

Addressing Thaetetus Socrates asks:

> SOCRATES: Once more we shall have to begin, and ask 'What is knowledge?' and yet, Theaetetus, what are we about to do?
>
> THEAETETUS: Regarding what?
>
> SOCRATES: Like an unbred cock, without having won the victory, we walk away from the argument and crow.
>
> THEAETETUS: How do you mean?
>
> SOCRATES: After the manner of antiloquists (Lys; Phaedo; Republic), we were satisfied with mere verbal consistency, and were well pleased if in this way we could win the argument. Although professing not to be mere polemists, but philosophers, I suspect that we have unconsciously fallen into the error of that dreadful class of persons.

Also for Feyerabend the question is not solvable, the choice of a research program is after all a bet; "sciences have no common structure; there are no elements that occur in every scientific investigation but missing in other fields (...) investigation does not owe its successes, when it obtains them, to obeying general criteria; it is based now on one rule, now on another and the moves making it advance are not always explicitly known" (Feyerabend 1984)

Duhe, Mach, Boltzmann, Einstein and Wittgenstein are of this line of thought; like a lucidly delirious gang they support the idea that a scientist more closely resembles a shameless "opportunist" than a "dutiful worker" devoted to the absolute laws treasured by the high priests of objective truth.

Data are complex because they are not immediate, univocal and sensitive as in naïve positivism (not for Comte), because they are not malleable nor subject to regimentation in a disciplined perspective as in Popper's falsifiability.

Answering a letter of Saul Rosenzweig's, Freud is lapidary. Anticipating again the question of data complexity, he criticized the psychologist who was searching for evidence in favor of the experimental verification of repression, expressing himself through these unequivocal words: "...I have with interest examined your experimental studies on the scientific validity of psychoanalytic assertions. I cannot set much value on these confirmations, as the abundance of reliable observations on which these observations lie makes them independent of empirical verification" (MacKinnon 1964). For all that, Freud does not deny the connection between theory and empirical knowledge; rather he criticizes the experimental method itself which seems to disregard the infinite variables implicit in the complex patient-analyst system.

Chapter 8

CAUSE AND WORDS

Freud seems to underline the inadequacy of non-clinical trials, a question which Groumbaum has recently posed again (Migone 1995). He would deserve more attention from clinicians' as he is one of the few philosophers of science who grasps the complexity of analytical research objectives, undertaking to defend psychoanalysis from hermeneutic drift and nailing it down to its responsibility as a natural science, even if it has to do with the domain of meanings and words.

He thus clashes with the positions of Habermas from the Frankfurt school and Paul Ricoeur from France. These two authors resumed Dilthey's distinction between spiritual science and natural science and placed psychoanalysis inside the first as it does not respect the conditions of proof, repetition and confutation. The only validation of psychoanalysis would thus lie in the psychoanalytical novel, whose creativity is based on inner narrative coherence.

Narration would therefore be the only validity of a discipline which is excluded from the assembly of natural sciences.

It is here that Groumbaum mounts his severest criticism justly handing over to psychoanalysis its epistemological responsibilities. It is not a matter of deciding what is natural science and what is not, it is a matter of understanding that natural science is a human science no less than human sciences are natural sciences.

Let us see how Groumbaum argues against hermeneutic principles.

His criticism of Habermas is severe and deals with the question of psychic causality which would, according to Groumbaum, be comparable to the conception of the natural sciences.

Freud states: "Human spirit and soul are objects of scientific research just like any other thing which is extraneous to man. Psychoanalysis has the special right to be representative of a scientific vision of the world (...) including investigation of human intellectual and emotive functions in science, (...) nothing changes in science's global attitude, no new forms of knowledge or new research methods result" (Freud 1932).

Hermeneutics therefore proposes a new perspective, of which Freud had never thought, as he on the contrary, conceived the scientificity of the clinical theory "on the strength of a secure and direct epistemic warrant from the observation he made of his patients and of himself. In brief, during all but the first few years of his career, Freud's criterion of scientificity was methodological" (Groumbaum 1984).

Natural science does not necessarily mean a science which is reduced and cleansed of the historical dimensions innate in the theoretic dimension.

Therefore also H.G. Gadamer is wrong, as he suggests an objectification of the experience, a cleansing of its historical component.

Habermas and Gadamer caused an untenable rift between nomothetic sciences and human sciences.

Groumbaum confutes Habermas and Gadamer's arguments on the basis of the exclusion of the historical element, underlining that both phenomena studied by electrodynamics and those dealt with by psychoanalysis imply a total dependence of every element from the context and from individual history, whether it is an energetic charged particle or a dynamic thought.

For Groumbaum a patient cannot have the privilege to cognitively validate or discredit psychoanalytic hypotheses; they are an internal component of the empirical verification of the interpretations concerning the patient himself, but they are not sufficient, as Habermas himself indirectly suggests. An intersubjective validation between patient and analyst actually proves to be necessary, beyond the validation methodology requested by Habermas, which is naïve as it is self-referential to the same observer, that is to say the patient, should objectively tend to the truth of an interpretation. Also Thouma and Kachele similarly contest Habermas's conclusions, underlining that it was precisely psychoanalysis which suggested the importance of distortions related to the simple and univocal introspection.

Paul Ricoeur's point of view is criticizable, too: he thinks that psychoanalysis is relevant only due to verbal exchanges between analyst and patient in the clinical transaction.

It is not possible however to exclude non-verbal exchanges from analysis.

On the contrary, non-verbal contents are the source of many fundamental discoveries of an analysis; they are more evidential proof than heuristic evidence (Jones 1959).

Groumbaum thinks, for example, that dreams are not only verbal statements explained during wakefulness but real facts pre-existing their verbalization. It is true that dream reports have linguistic gaps which substantiate a dream. Groumbaum uses this metaphor to better explain his point of view: "It is salutary in this context to appreciate that what astronomers endeavor to explain, in the first instance, is not the visual impressions they have of celestial occurrences, but the celestial events *retrodicted* from these impressions, or the events whose earlier occurrence their impressions represent. And just as dream reporting may well be inaccurate, so also the accuracy of astronomical observations has been lessened by the earth's atmosphere, for example. Small wonder, therefore, that what Freudian theories try to explain, in the first instance, is not the verbal dream-*reports* made by dreamers to analysts or others, but rather the manifest dream-contents inferred from the dreamers' subsequent mnemonic impressions" (op. cit.)

Here the analogy has evidential value; it is used by a universal explorer of phenomena which are not reductively delimited in compartmentalized and regimented fields by an improbable, unique methodological specificity, solitary and absolute, of the investigated processes.

This is complex reasoning too.

Anarchy and method support each other, create new scenarios promoted by their paradoxical relating, antinomy creeps in the knowledge object and in knowledge itself as an object, closing a loop which envelopes all that can be considered a concept with no discrete solutions.

From an epistemological point of view Paul Ricoeur is naïve enough not to consider fantasies of seduction causal (or at least co-causal) factors, just because they are fantasies and not events that really happened. For psychic determination there is no difference if the cause is a real trauma or if it belongs to an autonomous product of the mind.

Ray Schafer also (Schafer 1976) makes the same mistake. I think Groumbaum too hastily gets out of this impasse, suggesting a validation of psychoanalytical concepts, through non-clinical tools; in this way his complex conception of the datum fails and is submitted to be evaluated and justified through a linear deterministic experimental perspective.

The conclusion is not completely acceptable because it is in contrast with the concept of over-determination in the acceptation of multiple causality as it is used in quantum mechanics.

Interactions between multiple variables do not result in a unique determination of effects, hence experimental validation lacks at least the principle of replicability, unless the validation itself takes place on the basis of very general statements and therefore far from the detailed discoveries emerging from the observation of a single case.

This may be valid at the level of wide nosological categories compared to rough causal factors; this, moreover, reflects on generalizations which imply a probabilistic prediction with an excessive degree of uncertainty.

In psychoanalytic theory unconscious ideation, as it is psychic and non material, does not deny the causal validity of ideation itself.

Now the problem is clear; we need only understand what kind of determinism and causes are involved, if they are efficient causes or highly systemic data complexes.

B.B. Rubinstein also thinks that an explanation in terms of causal relations, including the area of meanings is possible. He, as Groumbaum reminds us, has criticized hermeneuts as far as the oneiric theory is concerned. The fact that the concept of causality is of a mostly psychological and not logical nature does not put its operative validity in doubt, it remains a legitimate and flexible conceptual tool, in common with both human and physical sciences; its conceptual versatility indeed lies in this extra-logical nature.

This is what Rubinstein has to say about it: "For an interpreter, (for example an analyst) meaning is also that which originated the symbol, hence *caused* (or contributed to causing) the occurrence of the symbol itself, otherwise it could not be said that the "symbol" signifies the meaning.

We will see that if we advisedly say that a symbol means something, then we can also say that it *represents* this something."

Some hermeneutic psychoanalysts also suspect the existence of unconscious causality; as Groumbaum again reminds us, they casually use cause-effect relationship seemingly without being aware of it. Thus they categorically exclude the possibility of a hermeneutic explanation referring unwittingly to the Freudian model.

For example Gould and Shotter openly discuss (unconscious) meanings which actions had for those who did them, referring, through the past tense, to a hypothesis of significance, which is certainly not constructed hermeneutically in a therapy session.

Therefore credit is given to Freud's position and to a psychoanalysis justly rehabilitated with the consensus of post-neo-positivist disciplines to the detriment of that creeping nihilism which tends to disprove everything.

In accord with Groumbaum "I believe that the generic disavowal of causal attributions clamored by radical hermeneutics is a nihilist and also frivolous trivialization of Freud's entire clinical theory.

Far from being a new bulwark of psychoanalytic apologetics, these hermeneutics, I am certain, are the kiss of death for that heritage which should have been saved" (op. cit.).

Repressed motivations are therefore causes of thought and behavior. Also in Freud's argumentation the term cause is used in its etymological acceptation as the motivating factor to action.

And yet Schafer insists in maintaining that motivations, being characteristics of the personal meanings and purposes of the agent, cannot be causes. Reductionist physicalism arises again, according to which states of mind are considered epiphenomena with no causal relevance.

I completely agree with Groumbaum's theories up to here. I start to be doubtful when he proposes the experimental nonclinical method as the only valid method of scientific comparison and controllability as if it were superior to clinical validation; which, above all maintains contextual conditions which are faithful to reality, by definition. I do not therefore agree even with Hans Eysenck who claims that "we can no more test Freudian hypotheses on the 'couch' than we can adjudicate between the rival hypotheses of Newton and Einstein by going to sleep under an apple tree" (Eysenck 1961).

My position is more similar to Glymour's: the only validation necessarily comes through the study of live clinical interaction tied to the specificity of each case (Glymour 1974).

I add that it is possible to apply a methodology of empirical description which is expressed in many *dialects* which are not necessarily and strictly clinical; for instance an analytical transformation can be expressed through symbolic-mathematical languages, like those used by Second-Order Cybernetic theorists in the description of *transformations* of data flows.

Even though in the clinic analytical language has a high heuristic value, we cannot exclude that other descriptive languages and perspectives could provide empirical validation and its *operational declination* with useful, though non-quantitative, tools.

Another long known example has come from a clinician and analyst: the Bionian *algebraic* language.

Chapter 9

IN THE BEGINNING WAS THE UNCANNY

Facts do not penetrate a world where beliefs live.

Paul Valery

Around the middle of the 19th century science was about to demolish the foundations on which it was based; the change was taking place in a silent and subtle way, nearly unbeknown to itself and on the basis of the developments of thermodynamics.

Till then Laplace's mechanism had prevailed; in *A Philosophical Essay on Probabilities* he had claimed that if an intelligence knew all forces and relationships between elements of the universe at a given instant it would be able to predict every evolution of any phenomenon contained in it.

Indeed, the decline of that mechanist perspective was due to the concept of energy proposed by thermodynamics.

Classical thermodynamics is based on two basic principles.

The first law states that nothing can be created or destroyed but everything is transformed; nothing therefore is ever lost but the whole is a big system of transformations.

Mechanical energy can be transformed into electrical energy and electrical energy can be transformed into chemical energy and so on.

Every system has an amount of internal energy which is the sum of the different forms of energy contained in it (from kinetic energy to the one which keeps its atoms coherent).

The amount of internal energy in thermodynamics is given by U and its absolute value cannot be specified; only its relative value cans, that is to say the difference between the value of energy of the same system before and after a transformation. One of the entailments of the first law is that the internal energy of a system can vary, but only if this system is not isolated, that is to say if it can exchange energy with its surroundings.

For classical thermodynamics, exchange of energy between a system and its surroundings can essentially take place through two *forms of energy in transit:* by work or by heat transfer.[1]

---

[1] The quantity of heat which is exchanged is conventionally written as Q and work is written as W. The first law of thermodynamics is therefore expressed by the relation $\Delta U = Q - W$, that is to say the variation of the internal energy of a system is equal to the difference between the amount of work and heat exchanged with the surroundings. The transfer from one state into another can occur in an indefinite number of modalities.

The entity of the energy transformation due to the transit of a system from one state to another is indicated by a *state function which* represents the internal energy of the system.

The second law exhaustively formulated by Clausius (1850) states that the heat energy transforms through degradation therefore it cannot completely change into useful work losing part of its capacity.

Carnot reached this conclusion by studying the transformation of heat into work; he imagined a cycle made by an *ideal heat engine* which takes heat, converts part of it and returns the remaining amount of heat into a cold reservoir.

Carnot observed that a perfect engine cannot be built and that a part of the heat is always returned.[2]

This is an irreversible degradation and it is called entropy. In a closed system, that is to say a system with no external energy supply, entropy can increase until it reaches a maximum at which it completely loses its capacity to produce work.

Statistical thermodynamics goes even one step further, Boltzmann identifies heat with the motion of particles: it follows that thermal energy is equivalent to the kinetic energy of particles and that an increase in entropy is actually thermal energy dispersal.

Entropy is a key concept due to its implications in the general conception of physical science, in the biological sciences and in every field involving the phenomenon of functional organization of matter.

Boltzmann's considerations, we have just seen, present the issue of entropy in terms of order and disorder, therefore in terms of organization and disorganization.

As scientific knowledge evolves, this concept proves to be a link between sciences which conceive entropy as energy degradation and those which consider it the reduction of order and organization: a new and contrasting idea causing important changes of perspective.

Laplace's determinism collapses, as it is no longer possible to predict transformations of a system in a univocal and accurate way. Thus knowledge can only be probabilistic and must be subject to contradictions. .

Moreover, whenever classical and statistical thermodynamics are proposed as a model of comprehension of large-scale phenomena, they lose much of their heuristic value; it is clear that the observation according to which the tendency toward disorder conceived as an irreversible fact and destined to increase up to a ultimate state of *heat death* is an abstraction which could not explain the organization of living and nonliving matter.

---

[2] The maximum efficiency of the ideal engine η is defined to be:

η=1-TC/TH, where TC is the absolute temperature of the cold reservoir and TH is the temperature of the hot reservoir.

As TC/TH is always greater than zero efficiency is always less than 1.

Therefore it is impossible to completely transform heat into work. The second law can also be expressed through the concept of entropy, a state function which represents the ratio between the differences of exchanged heat and temperature and it is represented by Δ= (Sb-Sa) = Q/T where Sb-Sa is the variation of entropy which reversibly transfers from state A to state B.

In Bolzmann's definition entropy is a measure of the number of possible microscopic states (or microstates) of a system in thermodynamic equilibrium, consistent with its macroscopic thermodynamic properties (or macro states).

The entropy S is defined as S= Kb log W

where Kb is Bolzmann's constant = 1.38062 x 10−23 joule/kelvin and

W is the number of microstates consistent with the given macro state.

Therefore, thermodynamics not only jeopardizes the certainties of determinism, but reveals a further innate complexity which makes the relationship between order and disorder but founded on logic and a dynamics which need further study.

The fundamental question, the essential paradox is based on this point: how is organization possible in open systems, exchanging energy and information with their surroundings? How do they keep it alive? And how does organization in nature complexify until it reaches the level of biological organization and even more that of psychological and social organization?

This seems even stranger considering the principle of a continuous dissolution of organization which, based on the above considerations could be defined as a principle of continuous destruction, sustained as we have seen, by the second law of thermodynamics.

At about the same time a similar process occurred regarding the study of phenomena in microphysics; the discontinuity of atomic order is discovered: its nature is quite different from the disorder revealed by thermodynamics, it supplies evidence for a nature of deeper levels of physical order and seems to bring into question the very concept of particles, , at the time recognized by Ruthefort as a sort of microcosm of the cosmic order.

The planetary system was the model which, reduced, should have made sense of the functional architecture of the micro-world which ordered matter at the level of particles. Particles vary depending on the observer, their nature can be either energetic or material; their coordinates in space and time are determined more by contingencies than univocal observational axioms.

The nature of elementary particles was defined on the bases of the quantum of energy, a key concept developed by Max Plank that re-established the nature of the discontinuity of matter, a nearly untouchable topic, as it forced theoretical research to equip itself with new instruments of mathematical and experimental analysis.

Therefore also from this perspective of the micro world the observed object looks more complex, multiply determined by different sophisticated aspects further complicated by variables related to the observer.

Disorder reigns supreme, even if it has a complex and different nature according to the various scale perspectives from which it is observed. Could this not be due to the effects of our limited powers of observation.

And again how can it be reconciled with order, which is the undeniable complement which paradoxically takes the features of the uncanny.

As there seems to be an indefinitely growing disorder, which does not prevent order to organize itself, on the contrary it seems to be accompanied by an indefinite and unpredictable growth of order, how can all this be explained?

What kind of universe are we speaking about, what kinds of knowledge are we expected to deal with, as it internally envisages paradox, antinomy, and discontinuity but simultaneously order and consistency? If order is accompanied by the growth of disorder (or vice versa) are they in some way connected? By what principle? What will become of the cosmos as we have always imagined it? How can we cope with this uncanny too? How is this going to reflect on the whole body of scientific principles? And on scientists?

The question is serious but, instead of letting it overcome us, it seems more useful to consider it a new interpretation of the real nature of complex phenomena, in their paradoxical becoming.

If were disorder – which was so intimately connected with order as to constitute the same source, contrary to what the second law of thermodynamics states?

This is what Ilya Prigogine suggested, claiming that there is a relation of complementarity and not necessarily of exclusion between order and disorder as the paradigm of simplification explains, excluding paradox and antinomy.

In this case order and disorder would be linked by a reciprocal causal relationship which reveals a phenomenal logic unusual for a linear and classifying reasoning.

Chapter 10

THE EMERGENCE OF ORDER

Boltzmann's contribution is important for various reasons, among which the fact that it permits disregarding heat in the experimental datum and operating in a more general context , which is precisely that concerning the concepts of order and disorder within a mathematical and informational acceptation .

Boltzmann considers natural processes in their statistical irreversibility according to which any disorganization of the particles produces further disorganization, as well as further increase in disorder.

According to the most recent developments in thermodynamics, statistical irreversibility does not only positively result in an increase of entropy, but also has a much more complex and *varied* destiny.

These considerations originate from experimental observations on physical and many other natural systems, which aroused suspicion that the surprises of natural reality, especially when one deviates from simplifying models, did not end with the passage to *probability models.*

The first consideration is introduced by Prigogine, who studies the so-called *Bénard vortices.*

By heating a certain amount of liquid, at first , various turbulences and highly non-organized phenomena will arise, but when the temperature is increased, at a certain point one can notice the formation of innumerable hexagonal cells which gather the ascending liquid by convection till they reach the surface.

Thus with increasing temperature, disorder has not increased; on the contrary it has become organized in hexagonal structures, that is to say in highly regular patterns.

Therefore, under certain conditions, order can originate from disorder; and what if this were true even for other phenomena regardless of experimental data

It is precisely what has been proven through mathematical computer simulations , where order and disorder have been recreated through numerical particles, subject to operational manipulations which have confirmed the expected result on the basis of empirical observation.

Henry Atlan has particularly concerned himself with this aspect; since 1971 he has developed numerical models proving that evolution of organized systems consists of "a succession of controlled disorganizations followed every time by restoration at a greater level of variety (*of the system's behavior)"* (Atlan 1971).[1]

Atlan has proposed an informational model, which permits understanding the logic of systems which organize themselves through disorder, that is to say chance effect coming from the environment.

These are systems which continuously update the informational program through which they subsist and continuously integrate disorder (chance) within parameters the system change according to the emerging characteristics; parameters beyond which they would be destroyed.

Dissipation (disorder) can therefore generate structures (order); and what if this phenomenon was a basic organizational principle of every natural process from molecules to the mind?

Natural systems are open systems, that is to say they live interacting with the environment. They describe their processes according to irreversible probabilistic models which cause the emergence of order.

Their evolution is highly unstable due to the unpredictable nature of their modifications. Prigogine calls them dissipative structures which belong to the domain of non-equilibrium and the possible at every moment.

Chapter 11

THE ARROW OF TIME

One can claim therefore that classical mechanistic science, based on Laplace's assumptions, is ruled by the paradigm of simplification; determinism is its law according to which certain causes are followed only by certain effects which are immutable and reproducible.

In this model, probability is nothing but an error related to our imperfect knowledge of the configuration of the studied object.

Many systems of the physical world function around equilibrium; for these systems (such as mechanical systems and many chemical reactions) casual variations can be ignored and deterministic considerations are still valid for them.

Let us consider, for example, a system where chemical reactions take place; it tends to proceed towards a state of chemical equilibrium based on particular concentrations of the given substances. The reactions are due to the nature of the reactants and to thermodynamic variables such as temperature and pressure. Equilibrium is given by the ratio between the concentrations of reactants and the constant value K, the equilibrium constant. This constant value represents the attractor toward which a system close to an equilibrium state tends.[3]

For dissipative structures, that is to say for systems which are far from equilibrium, changes due to their own functioning tend, instead, to amplify and to invade the whole system of which they are a product.

These amplifications are caused by minimal or infinitesimal variations of the original configuration and can generate irreversible states, that is, states that never come back to the initial state of order and disorder. Systems far from equilibrium are for this reason called dynamic and however precise our knowledge of "initial conditions is we can predict only the formation of one of the numerous possible structures": deterministic law is an artifact and probability is fundamental (Lazlo 1992).

Mathematical speaking it is said that processes which tend to equilibrium and order can be approximately described by making them ideally simple through linear equations, in other words, by making processes similar to "a system (...) whose behavior, due to multiple stimuli simultaneously acting on it, is equivalent to the sum of the behaviors derived from the separate stimulus"; on the contrary, processes which are far from equilibrium are described

---

[3] The attractor of a system of chemical reactions close to equilibrium can be modified by particular substances called catalysts. Thanks to them the constant K can vary generating new attractors. In the biological field this action is caused by enzymes, I will elucidate the concept of attractor later.

through systems of non-linear and non-integrable equations, systems with numerous possible solutions (Garella, Napoletano 1996).

Irreversibility and non-linearity of processes originates diversification, beauty and the complexity of the natural world, constituting the creative dynamics of very different systems: from biochemistry to meteorology, from cosmology to psychology.

And it is a dynamics which facilitates the autonomy of the systems which are now free to adapt themselves to changeable and unpredictable environmental variations.

All this involves a radical epistemological change which makes us observe the world in terms of uncertainty, and of the ephemeral but at the same time creative self-organization which discovers time as fundamental concept to the comprehension of the evolutive process.

Emergences, that is to say new organizations of matter and of its functions with their respective and different *levels of reality* do not seem so different; on the contrary they seem based on the same deep underlying logic which transversally informs them.

From this new perspective phenomena can be explained neither by deterministic causality of classical mechanics nor by Darwinian causality.

In a universe thus organized, not only the fittest survive thanks to the features which guarantee their survival; there exists also a creative and collaborative competition between the evolving systems.

Stability is temporary but it originates innovations which are useful both to the system and to the context from which the system emerges.

Chapter 12

OPEN SYSTEMS

The real innovation of our way of conceiving evolving systems is represented by the concept of dissipative structure s or open systems. This concept is indicative of an unlimited, unpredictable, non-linear process able to generate as it has been said, temporary states of either order or disorder.

These states, in turn, thanks to interaction with the environmental variables with which the system interacts, can lead to further upheavals causing new, equally unstable organizations.[4]

This leads to question what variables, once introduced into the systems, can destabilize order and generate a creative instability.

And even if we could answer this fundamental question, the change brought about would in any case be highly specific and therefore we could not know the result of our action on systems which follow similar general laws.

The above-mentioned experiment of Bernard's cells is one of the reproducible examples, a sort of chance event, but in nature things work in a different way.

Inorganic chemistry also provides an example: it is a state emerging from a highly nonlinear process, reproducible as well within certain parameters which do not yet exist in that turbulence.

It is the Belousov-Zhabotinsky reaction, consisting of oxidation of an organic acid by potassium bromate in the presence of a suitable catalyst.

---

[4] Open systems also can be defined through the concept of entropy. This is exactly what Ilya Prigogine does, giving a rigorous mathematical definition.

dS describes the total variation of entropy in a system open to energetic and informative interactions with the environment, diS describes the result of the sum of the variation of entropy produced by irreversible processes internal to the system, deS represents the entropy entity which enters the system. Hence $dS=diS+deS$. In an ideally isolated system the value of dS is constantly positive as the work of the system is by definition identified with the increase of intrinsic organization, hence with the decrease of entropy or disorder. Things are different in an open system, as energy or information entering the system can increase or decrease the total entropy. An open system can be also in equilibrium zero if the algebraic sum of diS and deS is equal to zero; in this case the system is in a steady state.

What is surprising also from the mathematical point of view is that in nature, the total entropy can decrease and therefore order can increase, violating the second law of thermodynamics, at least for local phenomena, the variation can be less then zero, that is to say $dS<0$. In this case we obtain $deS=(-diS<o=0)$ that is to say entropy is transferred into the system. (Cfr. E. e C. Lazlo, Franco Angeli, Milano, 1994)

The whole system can be seen to act in unison seeming able to make the involved molecules communicate at a distance. . The molecules behave in a synchronous and highly organized way alternatively and massively changing their systemic behavior.

There are other spectacular examples which demonstrate phenomena of open systems with a nonlinear, highly unpredictable and consistent behavior, such as Winfree's experiments to which I refer as to Zhabotinsky's (Pedone 1995).

Another example of a nonlinear process comes from biology. It was studied by Goldbeter and Segel; they analyzed the development of the aggregation system of acrasial amoebae, a kind of mucilage mildew (Dictyostelium discoideum) representing a sort of conjunction between unicellular and multicellular organisms.

The aggregation of many single cells leads to a mass, a pseudo- plasmodium, which soon develops a peduncle with the precise function of spreading the cellular material of which it is made in order to generate other similar proto-organisms.

This aggregative phenomenon is extraordinary because it uses cyclic AMP present in the environment or medium for amoebae and through this allows the propagation of the organizational nonlinear factor.

In essence it is a good example of order created through fluctuation due to the instability arisen in the medium where a *creative competition* is activated between the systems which evolve together; in fact also the environment (medium) is subjected to changes and reorganizes itself.

In the thermodynamics of far-from-equilibrium systems the process of the dynamics of emergent organizations and similarly of *catastrophic changes* entailing turbulence, is not regular and continuous but it assumes a typical discontinuous form, comprising sudden leaps and reorganizations.

There are precise energy points involving discontinuity and change, they are the so-called *bifurcation points,* which continuously pose the question: which direction will the system take after the bifurcation point? Here the process reveals all its instability, discontinuity, irreversibility and unpredictability; the universe of possibilities through which the system will be able to evolve depends on this magical point. The only certainty lies in in the fact that at bifurcation points the process very often evolves while complexifying that is increasing the level of order within its own intrinsic organization.

In natural systems there are multiple bifurcation points as the variables involved are innumerable; this is therefore the mathematical reason why prediction of successive states is highly improbable; nevertheless it is possible to identify some recurring characteristics in dynamic systems which function with a high level of non-linearity.

We will see in more detail that open systems are highly sensitive to the conditions of the system and to the smallest changes, which can indeed have remarkable effects on the whole system.

Moreover, as I have mentioned, open systems tend to ever higher levels of complexification, that is of organization which make them resistant to changes and at the same time more able to interact constructively (creatively) with their relevant environment.

Chapter 13

SENSITIVE DEPENDENCY ON INITIAL CONDITIONS

Until some decades ago scientists linked to Newtonian determinism (classical, Laplacian) believed that, given a certain knowledge of the initial conditions of a system, it was always possible to study its evolution even after long periods of time. Approximations are justifiable or rather necessary because they permit study of most physical phenomena.

Small influences can therefore be disregarded. If this is true for astronomical phenomena and for problems related to classical mechanics, it is not true as far as more complex sciences, such as meteorology or psychology is concerned.

Edward Lorenz, a meteorologist familiar with computers and mathematical theory, soon realized that prediction is not possible in physical sciences. Studying weather prediction through mathematical models, he discovered that the insertion of approximate numbers in a calculation flux causes divergence to increase more and more.

In other words a small, infinitesimal mistake in the computation of the variables introduced in the p calculation of prediction can generate a remarkable divergence in a very short time.

This phenomenon becomes visible when numbers are graphed. The reason for all this is the butterfly effect or sensitive dependence on initial conditions; the name was coined by Lorenz himself during a congress which took place in Washington in 1979.

At certain conditions of interaction of many variables a system is sensitive to initial conditions; hence mistakes and uncertainties become gigantic and break every predictive model.

This does not imply causality but aperiodicity and intrinsic unpredictability which can subvert .predictions.

In nature, actually, there are simple processes or complicated but simplifiable processes, which do not undergo the distortions of numerical approximation inserted in the equations which describe them. These processes are described by linear mathematical functions such as $y=rx$.

Moreover there are complex processes, which are not simplifiable and are caused by the interlacing of numerous variables; they are sensitive to minimal variations of the conditions introduced into the system and are mathematically described by systems of non-linear equations.

Chapter 14

COMPLEXIFICATION

Having exceeded the stability threshold induced by energy variation or information from the external environment, the system evolves with typical general characteristics, such as an increase in organization or of evolutive level; there is, in conclusion, a *development vector* which can be studied as a function of variables.

Despite the substantial unpredictability of these systems, it is possible to delineate some general principles.

To return to the development vector ; it can be defined as a function that over time is related to dissipation of entropy produced by the system, that is to the amount of order obtained and related to the number of components with a functional role in the system itself.

This function is even measurable in many physical systems while this is nearly impossible in high-complexity systems or in systems not easily quantifiable such as social, economical and psychological systems.

Nevertheless the heuristic, explicative value is worth considering; it can be explained in the following terms: the more complex a system is, the more it is able to use disorder in its environment or order of other systems which it interacts with, in order to generate further order internally; this process is always increasing, even if it can at times be forced to behave according to temporary inversions, leaps and evolutive discontinuities.[5]

In other words, systems increase their efficacy and their efficiency depending on how they interact with their environment, with evolutive but not evolutionary Darwinian methods.

Therefore, as I have underlined, what matters is not the characteristic of the fittest but rather the suitable characteristic which is co-created in the environment, with the collaboration of the environment variables.

According to this perspective nature is continuously searching for possible solutions in order to favorably impose itself within the organized order.

In this regard, open systems are assumed to have a natural tendency to complexify.

Another surprising aspect of open systems, or dissipative structures, as we are accustomed to calling them, is that during their development, they go through phases of relative stability and phases of dynamic instability or turbulence.

---

[5] In formal terms we can write that the development vector $S_{el/dt} \Rightarrow 0, (\Psi, I)$, that is the evolutive level of a system Sel in time dt has a positive, increasing value; Ψ describes the level of dissipation of entropy produced by the system and I describes the number of components functionally connected to the system.

This has a theoretical explanation which permits the comprehension of many aspects of the dominant linear deterministic vision of the world.

In the evolutive development there are indeed phases of relative stability, ruled by deterministic laws; that is, areas in which deterministic perspective permits us to understand the system dynamics to a good approximation because it is well-described and sufficiently precise.

There are also areas of high turbulence: at bifurcation points this approximation is no longer valid, it does not add up, there are multiple possible solutions, there is no stability, especially at a microscopic level; bifurcation *chooses futures organizations.*

This is why classical science and the paradigm of simplification have worked well however, despite the complexity of the real: they have simply scotomized everything that represented disorder, that which for some reason was not already known or not part of a predictable linear logic.

Dissipative structures have the merit of making us realize how vast the natural world really is and how we, more or less consciously, have adapted ourselves to our way of observing life; therefore a science of life has never existed, it was a science which observed life from the mechanical point of view.

An organism was at first considered a machine, then an incoercible error , sometimes a mysterious puzzle, now at last a system subject to observation and, if not measurable, at least in its deep logic, a logic which is more similar to artists' outlandishness than to the strictness of combinatorics.

Chapter 15

POSSIBILITIES

With the increase of complexification other typical phenomena observed in Zhabotinsky's experiments take place; the system seems to behave consistently on the whole, as if it were obeying a general law which completely informs it; it is an extraordinary phenomenon which has not really been explained yet.

It is even more interesting to note that it remains consistent while changing levels of scale at which the evolution of the phenomenon occurs and this has the most remarkable consequences at theoretical and applicative levels.

To return to bifurcation points, what happens when instability, turbulence is produced? First of all symmetry is broken, instability originates chaotic behavior in the whole system. While in the phase ruled by order there seems to be a communication at a distance between all the micro-components, hence the degree of freedom of every component is drastically reduced by the effect of general systemic order which cannot be reduced to the sum of the components; on the contrary in turbulence the degrees of freedom chaotically interact, transporting the entire system through improbable states.

Through a macroscopic approach, this can look like an ordered systemic behavior, because the behavior of a few single peripheral elements can affect the whole system, becoming dominant.

By breaking symmetry, disorder and turbulence are produced but also innovations and stability in an endless process.

It is something quite different from what occurs in phenomena of artificial intelligence in *expert systems* which do not behave as natural open systems, they are not dissipative structures.

Expert systems work via knowledge inserted in their operative memory by the knowledge of the system's programmer.

Dissipative structures, instead, self-organize by devouring environmental disorder or by systematically joining with other systems in evolution; their behavior is neither piloted nor programmed by anybody or anything.

A complex natural system acquires knowledge spontaneously and in its stable phase the new system's behaviors are completely invented anew, nothing predicted their arising in such an organized way until a short time before their formation.

There have been attempts, however, to imitate the behaviors of highly complex natural systems behaving like dissipative structures; they were mostly made by the Hungarian mathematician John Von Neumann who studied Cellular Automata (CA).

They are actually *simulations of virtual mathematical organisms* hence by introducing a repetitive rule, the system will evolve autonomously.

The studies on cellular automata were extremely important as they permitted to anticipate the logic through which genetic information was conveyed to entire living organisms and more recently laid the logical-mathematical foundations for other machines to be built. These systems, based on the so-called logic of neural networks, need to learn in order to solve problems (Pedone 1995).

On the basis of these recent evolutions in mathematical simulation it is possible to approach the logic of the living, that is, of the organic dissipative system. Although much has yet to be done in this direction, it promises to permit the creation of simulations of self-constructing artificial systems (autopoietic logic).

We will see where all this leads; fascinating is the idea of a spontaneous self-creation giving rise to systems whose implementation poses countless theoretical and practical questions to the natural intelligence which developed it.

So far only a few types of elementary logic suitable for producing artificial order may have been implemented; hence they are more likely to create systems which are similar but not quite identical to natural ones.

Let us leave the question unresolved as it would lead us to overshoot our target and also to deal with philosophical issues far from our intentions.

As Langton reminds us, one may think that a good reason to go on with this impracticable research lies in the possibility that systems which are similar to natural ones could contribute to the solution of scientific and *above all* technological, that is, mostly applied problems (Langton 1991).

By now it is clear that complex natural structures are beyond the principles of classical thermodynamics, energy and information do not simply tend to break down following the principle of entropy, but to continuously reorganize according to a principle of self-organization.

Chapter 16

TURBULENCES

Ultimately, what does turbulence mean? It is mingled with order; it is a very common phenomenon; it is the rule rather than the exception in nature.

Turbulence can also generate order under certain conditions, but are these really transitions from order to chaos in very different phenomena such as a meteorological event or a flow of capital due to stockjobbing? Moreover, why is the liquid-vapor phase transition similar to conductor -superconductor phase transition?

In all these phenomena transition to disorder occurs suddenly, resulting in new scenarios, with infinitesimal probability of repeating. Other analogies can be found between the organization of a termite colony and the chaos generated by the chance impact of a rocky mass on the termitarium and, likewise, between the pathological order of a psychotic condition, with its complement of positive symptoms and its split dynamics and the phase transition occurring in response to a change in the behavior of the entire medical team.

Is turbulence confusion and total randomness or does it hide a secret, and if so, what is it?

Theorists of complex phenomena argue that chaos is, contrary to popular belief, a condition of a particular organization, which can be studied through mathematical tools.

The so-called chaotic phenomena indeed prove to be highly and finely structured according to a quite particular symmetry.

Let us try to grasp what seems a real antinomy, a real contradiction of the deep logic of natural processes. As often happens, we have arrived here via mathematics, that is, through advanced computer simulations, certainly starting with the observation of phenomena followed by an attempt at virtual simulation.

Chapter 17

STRANGE ATTRACTORS

> *If* each city is *like* a *game* of *chess*, the day when I have learned the rules, I shall finally possess my empire, even *if* I shall never succeed in knowing *all* the cities it contains.
>
> Italo *Calvino*, Invisible Cities

It is also possible to visually represent what happens to an evolving complex system, with the passing of time, through a Cartesian coordinate system.

For instance, if the system consists of a simple pendulum, and if we imagine the pendulum in perfect condition that is without friction only two variables are needed, placed on the two axes of our suitably arranged geometrical space. The horizontal axis could represent the position of the pendulum's mass and the vertical axis its speed.

The system therefore will have a mass passing from an initial speed of zero to the maximum in the space defined by its amplitude.

The ideal system will geometrically (graphically) follow a circular trajectory in a mathematical space defined by the area delimited by the Cartesian axes.

In this case the attractor is said to be made up of a periodic motion circumference, where the system will be constrained to infinitely repeat its identical sequence in the phase space defined by the bidimensional axes.

Then, if we describe a real pendulum by introducing the variable friction, it is still possible to use a bi-dimensional phase space, but the attractor, as can easily be verified experimentally, is completely different.

Friction causes the swinging mass to slow down, and the pendulum's phase space will identify, after a certain time, *a point as equilibrium locus.*

We have therefore defined a static attractor which decides the final destiny of that system.

The phase space thus also permits visualizing the evolution of multidimensional states, by abstracting salient variables and drawing the combined trajectory in time.

Limit cycles and fixed points are only the simplest examples detectable in nature; why not consider similarly representing also multivariate systems?

It should be possible using some mathematical artifice; that is what Edward Lorenz did: he tried to represent a system of three interacting variables but he got neither a fixed-point attractor nor a limit-cycle attractor. Everything seemed to behave randomly until the three variables were put in relation in the space defined by those three variables.

He achieved something extraordinary, unexpected: the multi-determinant point drew a strange trajectory, with an elongated propeller shape in a three-dimensional phase-space. The repeating trajectories formed a typical pattern, always similar but never identical to itself, that is the moving points of the combining variables traced trajectories which were similar but never identical to those previous.

Here is the oddity, the paradox: how is it possible that there is an attractor wherein complex never passes through while remaining unmistakably itself?

Lorenz had found a stable but *aperiodic* attractor which was definitely strange: its invariance was made up of the geometrical (volumetric) space occupied, even if it was endlessly occupied by the trajectories of the states without being saturated; it was a simple but at the same time spectacular event.

An infinite number of trajectories are contained in a finite space; therefore disorder and turbulence indeed produce a finer order with particular characteristics, which cannot be identified through a common logic or simple common sense. Even though these trajectories are stably organized they are unpredictable, creative, neither repetitive nor typical. (Lorenz 1963).

Following Lorenz's experiments researchers identified various strange attractors, such as Birkhof and Shaw's funnel.

These considerations imply the need for great humility in dealing with the complexity of the phenomena under investigation. Experimenters, clinicians and those who introduce variables into highly complexity natural systems should be careful with these variables, which are only seemingly trivial or of little value.

Indeed, we should not so much focus our attention on macro-transformations of the systems, but on micro-events, just on those infinitesimal events which could be responsible for the great transformations concerned, that is the creation of new attractors; the question therefore concerns the *pertinence* of what has priority to be observed (Stengers 1988).

The aperiodic attractor immediately turned out to be a weird object due to its apparent logical absurdities, but that did not prevent the discovering of other numerous attractors in various fields, not even connected to the observed objects (Ruelle, Takens 1988).

The belief started therefore that there could be strange attractors organized in multidimensional phase spaces, that is, contained in phase spaces which had more dimensions than those obtainable through Euclidean geometry.

Euclidean space is actually homogeneous and isotropic. Chaotic phenomena are better explained via an Aristotelian concept of representative space which considers discontinuities, catastrophes or explosions of mathematical order both spatial and temporal, as well as collaborations and islands of stability.

Research revealed the presence of attractors in every scientific domain. In Japan, for instance, scientists studied strange attractors generated by electrical circuits, as did Jim Lesurf. Elsewhere they studied star orbits in galaxies in order to explain events which seem incompatible with common observations (Ueda1985), (Hènon 1964. Tom Mullin deals with the question of turbulence and non-linearity in fluid dynamics; Tim Palmer resumes the matter of unpredictability and the consequent difficulty of prediction in the study of weather; Robert May explores the biological field and finally Allan McRobie and Michael Thomson apply mathematical models to the study of catastrophes in Engineering.

The fields of particle physics and atomic physics are also included: Michael Berry studies chaotic phenomena and the possibility of evaluating their effects with a non-linear model in quantum physics.

Attractors can be studied also in human sciences; it is enough to mention Luc Ciompi's numerous works. Medical Director of the Psychiatric University Clinic in Berne, Switzerland, he tries to find attractors which identify and permit some predictions of relapses of psychotic decompensation.

In a recent work, he has tried also to apply other concepts related to the new epistemology of complexity, insisting, as we will see, on the non-linear multiplicative effect of transformations during the prodromal phase of an acute crisis.

In his work, published by the Italian magazine "Il Vaso di Pandora" in 1995, the researcher claimed that theories concerning chaotic phenomena would be very important to research on time and frequency distribution of psychotic decompensation (Ciompi 1995).

Psychoanalytical interpretations can also be conceived as instability generators in a system stabilized in a particular attractor field, "Through psychoanalysis, one can, in fortunate cases, reach the critical border zone where drive mysteriously takes a psychic shape, and one can influence the dynamic between chaotic drive processes and defensive structural orders by analytically disorganizing an existing order zone and patiently waiting for the following spontaneous reorganization".

Interpretation, then, implies putting a psychic system in a transition phase state where energy is essentially of an informational kind, verbally expressed; "on the other hand the information level tends increasingly to detach itself from the energetic-material level, as organization (for example, encephalization) increases, constituting a behavioral, instinctual and finally psychic domain, where the problem of equilibrium between survival and development , homeostasis and variation or better homeosis and change-tolerance is solved by gradually reducing the amount of energy" (Garella, Napolitano 1996).

According to my experience, the radical change in symptomatology achieved through a single interpretative event continued with a follow-up of several months, led me to think that I was facing an extremely important phenomenon of non-linearity, stabilizing the personology structure on different attractors (Lenti 1998).

Let us return to strange attractors and their impossibility of their Euclidean representation.

Mathematicians again solved the question, by bringing about a very simple change: they removed a geometrical dimension from the attractor responsible for the representation of the turbulence phenomenon under investigation.

They achieved the so-called Poincarè map, which is a transverse section of chaotic matter detected in phase space.

If the mathematical elaboration is *extreme* enough, in other words if it contains a sufficient quantity of data, quite precise and significant Poincarè maps can be produced. That is, sections which show progressively more detailed structures, until infinitesimal dimensions are reached and where we notice the same aperiodicity of the infinite set normally observed in phase space.

Apart from mathematical models, the concept of attractors seems significant, because it permits interpreting turbulence and catastrophic change, according to Wilfred Bion in the clinical field, as a change which can be hypothesized as, not random but highly organized according to nonlinear logic.

In psychodynamics, every defensive attitude, every psychic structure is not a static and repeatable condition but an attractive field where phenomena arise again, where previous experiences do not repeat identically, but comparably and invariantly.

The concept of attractor, therefore like the concept of non-linearity, must not be interpreted, strictly speaking, as a quantifier which restores numerical dignity to concepts, but a container of meaning which can convey contents of language, of emotions, of affectivity, in a word, of thought, while maintaining certain connatural characteristics, in particular unrepeatability and invariance.

The concept of attractor helps us configure the antinomy of complex states of psychic life without sacrificing either the redundancies or the apparently marginal deviances.

In complexity nothing is marginal, even a minor event can have major repercussions, both quantitative and qualitative, on emergent qualities.

Chapter 18

BOUNDLESS OBJECTS

Back to the question: what is a complex system? First of all it is something which cannot be gripped in the vice of an excessively rigid definition; otherwise it will lose an essential component of its nature.

What is complex is so called also because it can only approximately be described, it is *a boundless object*, as its description depends on the theoretical and methodological reference framework, on the variables under observation and on the context in which it is studied (Di Nuovo 1998).

Moreover, scientific certainties about objects are culturally determined and result from intersubjective observations and considerations (Cerruti, Lo Verso 1998).

This approximation of the object of study has repercussions also on the language which is used to describe it.

Psychoanalysis describes phenomena through common language which, as Bion reminds us, even if it is something more than a passive means of recording which allows us to work is much less precise than mathematical language through which an object is evidenced by physical quantification tied to measured physical values.

Nevertheless it seems impossible to describe psychic processes too. They are only partially quantitative, such that, as I see it, they are symbolic and qualitative processes which cannot be described in terms of elaboration and informative *transduction.*

It is, for example, the case for the concept of constant conjunction, introduced by Bion in Learning from Experience. It depicts the transformation into a symbol of a content, which, not having been evacuated, is represented in the first steps in the learning experience by conjunction of an "appetite" with an absence. Following repeated attempts at integration the content is elevated to a meaning (symbolic).

In Bion's original algebraic definition the conjunction is represented with an exponent n, which by joining container and contained, indicates the function being elaborated.

It is clear that the exponent n does not represent a number but a reiteration index of an informational process which, through numerous and partly mysterious steps, becomes a symbol.

It can be used, as a symbol and a sign, to indicate the concept of constant conjunction which is able to describe with acceptable precision, variously complex abstractions up to scientific deductive systems. It is difficult to detect the original connotations of container in the exponent n, but in fact they are contained within it, as it can be intuited.

The recursion indicated by index n is theoretically infinite, therefore Bion claims that content raised to the power of n corresponds to infinity; also in this case it is difficult to interpret this concept mathematically; it more adequately represents a system of reiterations and combinations of abstract concept with no prefixed limit.

In general, the elements on line H of Bion's grid are unlimited objects, partly because they can be specified only through every particular association via an actual realization.

Let us consider the concept of a psychoanalytic object; it is a function of a preconception identified only by a realization corresponding to a particular emotive experience.

Bion is explicit on this when, speaking about the concept of analytical objects, he confirms its boundlessness, as I see it: "Abstraction must serve as a constant in order to function as an unknown factor, but at the same time and thanks to the analytical object from where it moves, it also must have the qualities of a given *preconception*, unlike innate preconception, with a certain *(penumbra of meaning)"* (Bion, 1962).

Another factor makes psychoanalytic concepts inexhaustible, even if, despite this they are not operationalizable: the intrinsic relativity of definitions.

Let us briefly consider the concept of preconception, as intended by Bion; every condition which makes a realization possible originates from a preconception at any level of abstraction at which it may be intended; every condition is, therefore, a preconception for the next realization; so the value of the concept of preconception is absolutely relative.

A system constituted of many interacting elements whose behavior follows a non-linear logic can be considered complex. It is therefore a system whose following states are hardly predictable. With a discontinuous function it remains in stability zones even for long periods. It is generally sensitive to even infinitesimal variables, introduced by the experimenter.

Complex systems are very common; they represent a typical process of natural systems, especially at the highest levels of evolutions. Their organization is *transversal* as it is common to very different systems and organizations and it seems to be present mostly where interactions take place through symbols, as happens in language.

Therefore, not only human organizations are complex systems but also two people in a room, who are trying to understand the unconscious communication between them.

Complex systems open up new theoretical horizons for which it is no longer possible to think in simplistic (linear) terms, according to which certain causes are necessarily followed by certain effects.

Our everyday working tools, even the most common such as interpretation, clarification, comparison, containment and others are included in an epistemology which is quite different from that hypothesized by Freud.

If we do not revise the epistemological perspective, we will have to work like *idiots savants,* unaware of our theoretical and clinical scope at the risk of generating remote effects which influence the entire psychic system or emotional field in which we are involved.

For example why could a delayed interpretation of a projected identification cause a psychotic decompensation? What (inter psychic) effects May a patient's intra psychic scission have on the medical team's work? Why can a team regress to a level of base argument? Certainly there are explanations provided by different perspectives; complex thought theory may prove illuminating and supply interesting keys to solving problems.

Is it possible to increase efficacy and efficiency by acting in a complex system settled at a pathological level using a combined psychoanalytical and complex perspective?

Is it possible at this phase of theoretical and epistemological evolution of psychoanalytical work to speak in terms of convergent points of view?

Can the emphasis which modern psychoanalysis puts on transformative movements (field theory), typical of its new way of working, or on use of metaphor as a therapeutic tool, indicate an epistemological move towards complexity?

Indeed, these phenomena can be understood as based on the promotion of non-linear transformation. In other words on the use of that which most characteristically belongs to complex phenomena and their dynamics, that is, the way they evolve.

A dialogue between clinic and epistemology may be helpful in order to find, through common reflection, methods which can improve current analytical techniques or which can identify new techniques based on a new perspective; complexity is indeed an epistemological matter but not to be considered purely in philosophical terms. It has to do with procedural epistemology as it deals with the analysis of concrete processes and with the study of the possible concrete interventions on complex systems.

Chapter 19

COMPLICATED OR COMPLEX?

One must still *have* a *chaos inside* oneself to give birth to a *dancing star*.

Friedrich Nietzsche

We can imagine, therefore, an order which organizes itself through the turbulence, disorder and chaos of informational dissipation. We can imagine an order made up of relatively stable relations called interactions.

Our complex systems are unlimited *objects* as they do not have a physical substance; they are not geometrically symmetric nor are they virtual; they are real but they are describable through infinite perspectives; a certain grade of uncertainty and inexpressibility still remains in their inner nature.

They can change while keeping invariant , that is, continuing being themselves or rather, the more they change to adapt themselves to the different conditions of the medium on which they depend and which they transcend , the more they become themselves.

The constituent interactions are identifiable as elastic connections, they contort while still maintaining their identities; consider, for instance the analyst-patient complex system, in the case of a borderline patient.

Interpretations of the defenses based on splitting can be expressed through virtually infinite modalities; the integration slowly emerging from the therapeutic process leads to even very different configurations depending on the quantity and typology of variables; yet something will long remain identifiable in the transformational processes; splitting still taking place will be the typical splitting for that patient, it will even cause a variety of diametrically opposite countertransference reactions at different moments.

Nonetheless, we will focus on that splitting and not another one while we will understand, when it happens, that the splitting will be transformed thanks to our ability to recognize it.

It is a complex, circular and infinite game, where objects' nature depends on that particular unrepeatable combination of the factors at play, whether they are islands of stability or catastrophic transformations.

Boundless objects create (look for) bounds or borders , this way they can continue existing because a border is a limit but at the same time an opportunity for evolution.

It is possible to reach stability through bounds; they are the condition for combating disorder.

Order and disorder cooperate to create organization and bounds, the boundless (evolving) object born from this extraordinary, complex dialectic counters the second law of thermodynamics, according to which nothing complex ultimately could exist and evolve.

It is not a competition or a hierarchy of opposing natures in distinct processes; it is instead a *heterarchy,* a collaboration, a cooperation indeed between different interactional levels, which are however a part of the same evolutionary process.

The unity of boundless objects envisages opposites as parts of the whole, unity has to be multi-faceted.

In any case it is not possible to speak about order in the singular; there is not only one sole order, There are different forms of stability called order: from cyclicity to hierarchical order, from the data flow of a computer program to the symmetry of a mussel. Of all types of order we find most interesting that which is complex, evolutional, interactional and sensitive to intentionally devised variables.

Many years ago Angyal first defined the nature of complex interactional systems as unitary and multiple (Unitas Multiplex).

From a certain point of view these systems can be even be defined as multiple unities, partly because they are unitary systems if considered on the whole, while they are anarchist systems if they are considered from the perspective of their elementary constituents.

In conclusion, the whole and its parts are reciprocally irreducible and defining.

It is an antinomy, a puzzle and a paradox which has little to do with the common sense on which natural scientists base their research.

Perhaps that is the reason why scientists have insisted on trying to simplify, englobe and grasp what is complex in its entirety.

The Gestalt theory has functioned this way: weary of not getting to the essence of the object by dissecting it, as natural science used to do, it started to study it taken as a whole.

The Gestalt theory has simply explored the other side of reductionism: if we dissect an art work we do not recognize it; we get pigments which are a different thing as they have other properties, but if we observe the pigments as a unified whole we again see the original work.

The problem concerning the object however remains: is it then reducible, holistic, and globalizable?

Is it a purely formal concept as systems theorists sustained? Or does something elude us? A system is complex because it is multi-faceted: it must be simultaneously grasped as a whole and as a highly organized interactional system.

It has not been easy to move away from logic of contraposition considering only one aspect of the question, from a disjunctive logic, according to which an object is bounded, that is objectifiable and observable from a single possible point of view.

But complex logic is different: it suggests an inclusive paradox; complex does not trivially mean complicated.

In this case bounds and emergences create new evolutionary possibilities and at the same time preclude others that we will never know.

Let us see a clinical case which helps us clearly understand that constraints and possibilities are intrinsic aspects of the unique emergent result.

Luisa turns up at the first session with a friend; she is forty but she looks younger, she has a straightforward manner, with a nice smile.

After she has sat down, her behavior changes: she willingly speaks about herself but in a decidedly sad tone. She says she is suffering very much and for that reason she is being pharmacologically treated by a neurologist, who gives her antidepressants and tranquilizers.

She wants to try psychotherapy, feeling that her psychological problems cannot be satisfactorily solved through her current treatment.

Luisa is a widow, involved in a Catholic movement where she lives; she writes a lot about herself solely to keep herself company and sometimes shares her writings with some of her female companions from the movement. She can write well even if she doesn't have a high-school diploma, she shows a remarkable introspective ability and a particular motivation to remember what she experienced during a quite difficult childhood: her parents were often absent, being busy at work especially after opening a pastry shop.

She describes her father as a coarse, violent man; he died some years ago due to alcohol related problems; she feels her mother as submissive to her father, frightened of his reactions and unable to assert herself in her children's upbringing. Luisa has got a brother, younger by a few years, and a sister 14 years younger.

In the first period of treatment, she claims to have been the subject of her father's sexual attention, after her younger sister's birth; she reports an act of masturbation which remained isolated but changed her relationship with her father who, after that, showed himself more and more pathologically attached to his daughter,. He took the liberty of jokingly touching her bottom, and proved jealous of his daughter's male friendships during her adolescence.

Luisa feels she is her father's accomplice, inducing him to that act; she tries to remember that night more and more precisely in order to relieve her guilt.

We choose a biweekly vis-à-vis therapy, Luisa cannot stand the idea of the couch at sight of which she is horrified, defining it a torture instrument. The connection to the bed where she experienced her father's violence is evident but I avoid interpreting.

Luisa starts writing to me, instead of writing to herself between sessions; this helps her gather her thoughts and get ready for our meeting.

As therapy proceeds, her defenses turn out to be organized on primitive idealization, splitting, and protective identification. She is attracted but also terrified by a horror/sado-masochistic film she remembers seeing with her husband. She is frightened by the presence of the devil in this film but she also seems attracted to it. She naturally has not got a good relationship with sexuality: she often fantasizes about a man whipping her bare bottom with a belt; only this way she can fall asleep after masturbating; sometimes she unsuccessfully tries to resist these fantasies; sometimes they turn into a real self-aggression: she beats herself with a slipper or a belt until she feels intense physical pain.

I suggest that, by doing so, she traumatically repeats her father's abuses and thus, although suffering, feels loved by him.

She agrees and also starts writing me letters in which she implores me to stay close to her, to help her remember more of that night. I reply that, through the letters, she is trying to bridge the gap between us, in order to have me as a good parent always at her disposal.

She repeatedly asks me to enter her, and does so with an insistence that makes me think that on one hand she wants to introject me as a good object which can help her to contain her anxieties and on the other hand that I could help her though a fantasized penetration.

Luisa is probably identifying me with her father, and I interpret it so since erotization is the only means by which she can establish an empathic rapport with him. She soon begins calling me daddy, with a nearly sensuous pleasure; she starts her letters with "dear good daddy", thus indicating that she considers me the good object and projects her negative feelings onto her real father.

Luisa seems to express herself on several levels: one level, which I would call hysterical - histrionic, is marked by an erotized relationship with her own internal objects and her dramatized request for help; the other, more clearly pathological; where the request is on a more archaic level still with erotization but with a different connotation. This is a sort of protective barrier against fragmentation anxiety which can be mitigated by bridging the distance by any means necessary, even a masochistically erotized expedient.

As Vassallo Torrigiani states: "the nostalgic desire for a merged union in a single body with the mother *(or with an undifferentiated parent) (my* italics,), especially when it is upheld by an anxiety of losing physical and individual sense of self, is phantasmically prolonged in some subjects by a fantasy of one psyche for two people, of one body for two, of one gender for two.

The last fantasy, that of one gender for two people, would express itself in neo-sexual practices (Joyce McDougall's expression), which would meet the psychic need of extending the illusion of continuity with the other's body, appropriated via sexual fantasy."

Like in Orlan's body, the extreme artist described by Torrigiani, violence "ambiguously takes the form of perverse sexual urges as if erotization should help transform a dramatic and mortifying experience into a success, and restore a sense of self and a severely damaged narcissistic structure This mechanism, however, works only temporarily and even increases the initial need which one then tries to satisfy, using like a drug , the same strategies for containing the anxiety" (Vassallo, Torrigiani 2001).

Luisa seems to live in a state of "paranoid anxiety (with regards to her own perversion), whose essential quality, paralysis, leaves her incapable of action. There is no escape from the object of terror inasmuch as it comprises dead objects" (…), which can be revitalized only by "the healing capacity of the internal parents so long as this is prevented by neither oedipal jealousy nor destructive envy."

When dependence on the reparative capacity of the internal objects is prevented by oedipal jealousy and/or destructive envy, this restoration cannot occur during the course of sleep and dreaming.

As Meltzer remind us, "only an object of external reality which bears the transference significance of the mother's breast at infantile levels can accomplish the task. (...) .

When dependence on good internal objects is rendered impossible by damaging masturbatory attacks and where dependence on a good external object is unavailable or not acknowledged, an addictive relationship with a bad part of the self, the submission to tyranny, takes place.

An illusion of safety is promulgated by the omniscience of the destructive part and perpetuated by the sense of omnipotence generated by the perversion or the addictive activity involved (...)

Until such a narcissistic organization is dismantled and a rebellion against the tyranny of the bad part is mounted, progress into the thresholds of the depressive position is impossible.(...) The dread felt in relation to the tyrant is fundamentally the dread of losing the illusory protection against the terror and may be seen to appear especially at times when rebellion has been undertaken in alliance with good objects which are felt to be inadequate or unavailable, as during analytic holiday breaks (Meltzer 1973).

This is happening in our analytic relationship but, despite difficulties, interpretations are progressively more easily assimilated together with a good object which is less and less precarious. This goal may have been achieved as I was able to find the best balance between the aspiration to understand the observed phenomena and the ability to keep and assimilate them, with no need to rush their return, as K. Abraham reminds me.

The self-harming behaviors have disappeared and a more collaborative relationship seems to have appeared, even though dependence is still active.

I wonder what it would have happened if I had kept a more neutral mood and if I had not understood the necessity of a vis-a-vis analytical relationship. What bounds and new emergences would have arisen if I had not considered the daily letters written by Luisa as implicit transitional objects?

I think that without these technical parameters, I would not have been able to establish a relationship based on the introjection of the good object, despite the fact that the patient expressed the wish I could go into her. I thus precluded other possible developments (due to the complex perspective). Perhaps reaching a neurotic level of functioning would be impossible (constraint) from such an archaic relational structure may prevent a nevertheless something has been done and Luisa can, for now, go beyond the sado-masochistic level.

Complicated ,may admit, as a problem, one or no solution; it can be reduced to an algorithm, that is to a more or less intricate flow of decisions, it can also be isolated, taken apart and put back together, identically. It does have rules and restrictions, but these are predefined and can be modified through an active operation.

A chess match is complicated, although simply lifeless. A living system is endless, as it creates new bounds, it changes the game rules while the game is being played in a *complex* (natural) way.

A complex system is certainly complicated, but the contrary is not true: a complicated system can be not complex at all.

One of the bases of Gestalt theory relies on the indisputable principle that the whole is greater than the sum of its parts. This principle is absolutely true and verifiable.

Four random lines on a page are just a scribble, if you juxtapose them according to a certain order a pattern emerges (percept).

The axiom that the whole is greater than its parts is confirmed also at the level of cognitive organization.

Kohler made remarkable observations on anthropoid monkeys' ability to solve cognitively complex problems, which testifies for the emergence of a whole with new characteristics.

However, this is partial: holistic emergences are certainly greater than the sum of their parts, nevertheless this does not exclude the contrary: no less true is the axiom that the whole is smaller than the sum of its parts

Indeed, as already observed, the emergence of a complex system creates new qualities, new stabilities but also new limits which reduce the degrees of freedom of a system.

An organization, due to the motive of its formation, can evolve in many ways but at the same time this will exclude the possibility of evolving in infinity of other ways.

The parts each considered as a system can serve a certain number of functions. The whole, due to emerging constraints, can serve other functions but not many of those served by its various parts; in this sense the whole is smaller than the sum of its parts.

Every system determines *specializations* and evolutive possibilities but excludes others.

An example taken from genetics is the adjustment made by a particular complex molecule which inhibits the activity of a specific gene, thanks to a function of repression (constraint) caused by the emergence occurring at the level of complexity of genetic reproduction of life .

At a totally different level, another example is provided by the rules and technical parameters of the analytic *setting*; which can be considered constraints of the whole on the parts (emergences), caused by the coupling of two hyper-complex systems represented by flows (processes) of conscious and unconscious communication between analyst and patient.

The proposed analogical leap is not only legitimate but also scientifically necessary when we talk of complex systems.

A complex system, considered as a whole, is therefore something smaller and different from the sum of its interactions, and that being said, the quantification of a complex system is impossible.

Indeed it is neither the result of a calculation (that is a summation) nor a combinatorial, as it is always an unpredictable and unexpected result.

Chapter 20

New Organizational Equilibria

Our tendency toward a geometrizing perfection (...) is never separate in the artist from the evolution of the other about whom nothing can be clearly said.

Luigi Cavallaro

We have argued how in complex systems, states of order and disorder, chaotic and stable routines, evolve and how bounds and catastrophic fractures follow one another endlessly.

All these dynamics give an idea of the, so to say, explosive transformations typical of complex systems, as if dissociative forces prevailed against conditions of stability.

We should nevertheless remember, as has been discussed in the chapter dealing with complexification that the evolutionary level of a system increases over time.

Therefore, the new equilibria are actually supported by a growing organization, that is, by an increase of bounds but also of new possibilities knowledgeable about the emergent tasks of the organizational arrangement.

In other words, as Lupasco stated , "For a system to form or exist, its constituents, either by their nature or the laws that govern them, must be able, at the same time, to both attract and repel one another, associate and dissociate, to integrate and disintegrate" (Lupasco 1962).

While thermodynamics speaks about thermal death, disorder, homogenization or entropy, complex systems include these dynamics and go beyond them through the introduction of the concept of antagonism.

Complex systems organize and reorganize following an antagonistic logic according to which organization and disorder are inseparable phases of a unique, irreversible, dynamic and endless process through which life originates.

Antagonism is an observation point for the evolution of complex systems: a moment, a snapshot conveniently chosen to stop a stream of phenomena which would otherwise be incomprehensible.

Organization is the other observation point, the other moment, the other snapshot which fixes the form in a stable area as a momentary (relative) outcome of the process.

A complex system pervades itself, by generating, in a sense, its form and by reforming it again at a later subversive or catastrophic stage, as Bion states dealing with transformations.

Moreover, organization is, according to Thom, an *invariance* or an open dissipative structure which Prigogine considers, like any other open system, open to chance for organizing disorder (Thom, 1972).

The more complex this organization becomes the more it will be able to use antagonism and disorder in order to increase complexification.

Thus arises the contradiction, the order-disorder paradox as a new emergence, even if the contradiction lies in the modes of our knowledge and not in the phenomenon under observation.

Our failure to consider natural phenomena full of contradictions is a logical and psychological limit, "the world can show indissolubly complementary antagonisms which our reason translates into contradictions. It is a typical error of concepts which consider knowledge a mirror of reality, that reality contains contradictions which reason is designed to discover and record. Indeed, all knowledge is translation, and contradiction is the way in which our reason translates the black holes into which our logical coherence sinks. (...) And this enables us to see and comprehend the complexity of reality, which obliges us to think in a way which implies using contradictions and acknowledging uncertainties" (Morin, 1991).

Our thinking needs to take into account the paradoxical nature of complex phenomena so that Aristotelian axioms do not represent the last word on the complexity of reality. In order to think in terms of complexity we need to shift from an analytic thinking based on the Aristotelian law of excluded middle to a dialogic form of thinking in which "two opposite propositions are necessarily connected even though they oppose each other" (Morin 1991).

Freud also had already explained how paradoxes were typical of the unconscious; his discoveries anticipated the future discussion on the inadequacy of Aristotelian logic and suggested the necessity of expanding our thinking.

These arguments show that there is no point in either seeking explanations of the whole in its elements via scientistic reductionism or the explanation for the single elements in the properties of the whole, via a generic and global holism.

Dealing with complex systems needs to go beyond both approaches, which have always perverted the nature of multiple unities.

Reality is complex, and its complexity exceeds univocal definitions, in any way and from any point we observe it.

Moreover the whole, the complete complex system is able to *retroact on itself,* that is, to influence both its constituent parts and its totality.

This represents what I would call *holistic logic* which explains, from a complexity point of view, the structuring of many phenomena in the analytic field.

The whole (a system) can retroact on itself. Based on this assumption Morin claims that the whole is more than the whole and Tarski explains that no system is able to entirely observe itself; there are only different points of view, even the whole can observe itself from points of view.

There cannot be an absolute point of view, an explanation which can absolve itself by self-demonstration.

Therefore there is not even any conclusive, that is non-inclusive discussion on science.

In epistemology, that is in discussions on knowledge (episteme ἐπιστήμη), all that is furthest from the pontifical and judicial positions of the strict object sciences cannot help but call itself and become dialogic.

Chapter 21

THE SUBJECT-OBJECT

A realm lies there of forms to explore and harmonies to discover.

David Ruelle

If the mind is conceived on the basis of topological metaphors, as in the Freudian viewpoint, then it is to be treated as a reification, in which different regions and strata are inevitably found, according to a more or less defined spatial differentiation.

"Some of the more careful psychoanalytic theorists have taken pains to point out that the concept of psychic structure refers not to something substantive but to recurring patterns of experience and behavior over time. Yet, in common usage, deriving from Freud's manner of discussing psyche, as if it occupied spaces with structural properties, the spatial metaphor pervades psychoanalytic discourse on self." (Mitchell 1993)

A temporal metaphor, however, seems more suitable to represent psychic structure; "Selves are what people do and experience over time rather than things that exist someplace. Self refers to the subjective organization of meanings one creates as one moves through time, doing things, experiencing ideas and feelings including some self-reflective ideas and feelings about oneself" (Mitchell 1993).

Rapaport had already introduced a temporal dimension in the conceptualization of psychic structure and later Shafer contributed to a description where illustrative metaphors of psyche were first extended to a narrative dimension (Rapaport1957) (Shafer 1992).

As Shafer, indeed, suggests "A person defines situations and emotionally invests events with multiple meanings (...)

In this account reality is always mediated by narration. Far from being innocently encountered or discovered, it is created in an organized way" (Schafer 1983).

But only Mitchell conceived the self as the result of multiple narrations emerging from relational configurations which provide an absolutely dynamic conceptualization.

According to Bollas, as Mitchell reminds us, "the self of a person is the story of many inner relationships (...) There is no unified mental phenomenon which can be called the self. The concept of self should refer to the positions or points of view from which and through which we sense, feel, observe and reflect on separate and distinct experiences in our being. One crucial point of view derives from how others perceive us" (Bollas 1987).

We thus arrive at a concept of the psychic system, according to which there is no place for either a monolithic, static idea of the subject and of the object or for a sharp contraposition between subject and object of knowledge.

From this perspective, one does not conform to the naïve realism which eludes the subject of observation, considering the psychic system the only object; we could rather speak of systems which observe.

The objective is simply considered the event facing the observer (who, in turn, is subject to observation by the observed) but they are boundless and cannot be conclusively defined; moreover it is a construction of the observer, the subject who will be ready to conceive it based on their interpretative systems and their own verifiability criteria.

Is it a paradox? A logical absurdity? No it is complex reality.

Complexity seems, then, to open an epistemological glimmer on a field common to both the subjective and the objective.

Certain complex systems can be considered subject-object, to use a neologism whose meaning corresponds to what has been said about complexity in these discussions.

A science based on new postulates could be born; scientific fields, like psychoanalysis which from its beginnings has had to deal with complex objects, must become aware of the large number of new perspectives from which we can re-examine phenomena which have always been studied with legitimate but univocal and self-referential methods.

It would be interesting, for example, to study from the perspective of complexity, the phenomenon of projective identification which, with its numerous implications, constitutes by itself, a copious "multiverse".

Even though we thus complexify questions of concepts already examined, we do not complicate them, we view them instead in a different light.

Hence it becomes even clearer that, as there is no reifiable (simplifiable) object, neither is there a non-way of studying a complex system, the concept of complex being a fundamental concept which cannot be reduced to something more elementary.

There cannot be a simplified science as it would betray both subject and object; there can only be an expanding container of containers of models which uphold complementary and multivalued points of observation: *subjective-objective.*

The history of scientific ideas consists of a series of models, of ways of seeing the world, which are more or less useful to solve problems, and scientific evolution consists of an evolution of the scheme of things and reality; science develops through paradigmatic evolution (Khun 1962) (Di Maria, Giannone 1998).

There is no a single universe, we should speak of *multiverses,* possible and plausible to the same degree that the perspectives of observation show plausible.

In other words if we use a complex epistemology we set out on a path which is infinite by definition, with possible developments at every moment and in every direction and which is still subject to verification, analysis and criticism.

Chapter 22

FEEDBACK

The part in the whole and the whole in the part.

Edgar Morin

As I have just sustained, a complex system feeds back on its components and on its whole self.

We only need think of the infinity of feedback which a cell returns to its components, that is to its organic macro-molecules: proteins, carbohydrates, lipids of which it is composed. It indeed, as an organized totality, defines structures, quantity and functions, and these parts, in their turn, inform the cell (the complex totality), when to decrease or increase one ore more components.

Words also are parts of a discourse (totality) which feeds back on them, and they, in turn, are understood within the specific acceptation and structure of the discourse.

Therefore, from molecules to language, there is a similar *self-feedback,* which leads us to explore another dimension of complexity, the logic governing the relationship between the whole with its parts and between the parts with the whole.

Chapter 23

AUTOPOIESIS

Describing the evolution of a complex system over time, we cannot prescind from the organizational contribution given by the system itself at the zero moment, that is, from its own organization rules.

It can therefore be acknowledged that it is also a system based on a self-referential organizational system, in the sense that it guarantees its own production.

Consider for example a cell of an organism.

A number of processes can be identified within it: from the construction of proteins to the production of storage depots for sediments (such as vacuoles).

But the cell, comprising proteins, carbohydrates, lipids, water and more controls the construction (as a computing apparatus) of its own components in order to guarantee its persistence.

The computing apparatus of the cell becomes generative, transforming information (potential negentropy) into programs and strategies (organizational negentropy), which govern actions and phenomenal performance which being necessary to the existence of the generative apparatus, it contribute to the regeneration of the generator. Therefore starting from the computing apparatus, an endless cycle of transformation from generative procedure into phenomenal procedure and vice versa is generated and re-generated (Morin, 1980).

In conclusion a complex system is not only evolutive but also autopoietic, in other words it evolves through *self-reproduction.*

Or in other words, it is a closed system because it defines and produces itself and it is an open system because the environmental perturbations become part of the mechanism producing rules and organization.

How does this happen? What are the principles underlying this ordering?

It happens whenever every *functional unit* interacts by *coupling* with other units or with random (*chance*) and non-organized elements present in the environment of the system concerned (*medium*).

"Every time that the conduct of two or more units is such that there is a domain where the conduct of each unit is a function of that of the others, they are said to be coupled in that domain. The coupling results from the reciprocal modifications which the interacting units undergo during interactions without losing their identities. If they loose their identities during interactions, a new unit can be created, but there is no coupling. Generally speaking, however, the coupling gives rise also to a new unit which can exist in a different domain from that where coupled units maintain their identities. The way this happens, as well as the domain where the new unit originates, depends on the properties of the component units. Coupling is a frequent event in autopoietic systems, whatever their level of complexity" (Maturana, Varela 1980).

Therefore it applies to both biological systems and systems which are mediated by symbolic languages; coupling takes place according to similar criteria.

In general "coupling remains invariant, while coupled systems undergo structural changes, selected through coupling and therefore proportional to it" (Maturana, Varela 1980).

In order that structural coupling not cause degeneration of systems, asymmetry is necessary: higher-order (or presumably so) organizational systems, that is more organized systems, have to couple maintaining their own organization in order to make autopoietic lower-order systems.

The resulting autopoietic system applies that "pressure needed to produce higher-order autopoietic systems through coupling of lower-order autopoietic unities. It seems that the only limit to the formative process of higher-order autopoietic unities is the one imposed by the circumstances in which a unity can be specified in a given space" (Maturana1980).

It can also be seen in the process of structural psychoanalytic change that through systemic coupling both the complex unities find reason to change; however, if the setting rules are maintained, this does not entail a loss of identity, on the contrary, it is the maintenance of these rules that permits (in systemic and complex terms) an appropriate evolutive coupling which involves both the protagonists in the development of their respective identities.

Autopoietic logic can also be applied to the domain of the symbolic and then to the analytic field if we consider the processuality in this field as transformation of linguistic signs which can modify the likelihood of uncertainty and propagate codes that entail the special, distinctive practicabilities of a process, a field and its specific contents.

Empirical verification imposes a posteriori to admit autopoietic logic in a field where there are no discrete boundaries easily insertable during processing.

Here the models of comprehension indeed seem to be capable of wavering on the thrust of analogic proof while avoiding the crushing and pointless burden of delimiting signs (verbal and preverbal, pitches, sequences of sounds) as discrete elements more suited to processing.

Thus "no meaning is possible outside the field", the rules of the system-field "make the events and phenomena of the field comprehensible and communicable, as their validity depends on their common interaction and not on the single meaning which anyone can arbitrarily give them or on external referents. No meaning is possible outside a well-defined context" (Russo 1990).

Moreover, the autopoietic process is theoretically limitless; its outcomes are temporary and produce further outcomes.

In other words, "linguistic structures of symbolic levels operate in limited spaces which are defined from time to time by certain coupled elements present in the field and last only the time necessary for that couple at that moment. Other symbolic levels are ready to substitute the previous ones when the field organization permits" (Russo 1990 b).

In an autopoietic process the repertoire of symbolic levels or "possible worlds" in terms of modal logic, is theoretically endless in temporal development, as I have said before, and synchronically undecidable (unknowable) until they have been realized as there are no immutable and general laws of the field “but a sequence of developing choices, in part unconscious, which govern the relationship between chance and necessity” (Russo 1990 c).

So here the question of the pragmatic function of language is exhausted: as Wittgenstein has already claimed, there are no general and universal rules according to which only one language is valid for every conceivable context, as operational models are created each time on the basis of each specific communication relative to every new context.

A complex system goes beyond the concept of linear causality, it cannot simply entail input and output data processing and it cannot be described by first-order cybernetics, which Von Bertalanffy and Wiener developed, originating the general theory of systems.

First-order cybernetics deals indeed with circular logic, recursive loops; accordingly a cause influences an effect which feeds back on the cause, but the system is not creative; feedback loops can well explain stability in certain systems or certain complex organisms as far as particular functions and global stability , reached in some evolutive phases , are concerned .

Actually, an autopoietic complex system is more appropriately described by *circular recursive* logic, according to which in a complex system effects produce the causes by which they were originated .This logic is also suitable to describe autopoiesis, as we have defined from the organizational point of view.

This is a great epistemological leap which requires a new approach decidedly placing the observer as co-producer of the poiesis of the system.

The academic bases refers to the theoretical biology of the late 1960s and 1970s with authors like Humberto Maturana and Francisco Varela , Von Foerster 's second-order cybernetics and Terry Winograd's computer research.

It can also be stated that a complex system is autopoietic to the extent that the fundamental variable, comprising the network of processes (interactions) tends to be constant over time, that is, it tends to reproduce processes guaranteeing the identity (unity) or autonomy of the system.

“Therefore autopoietic organization simply means processes interlaced in the specific form of a network of productions of components which realizing the network which produced them constitute it a unity” (Maturana 1980).

These argumentations on processuality have the advantage of referring to a level of abstraction which transcends every particular realization.

This is a remarkable aspect as we can translate concepts into different domains and identify analogies in further observation domains.

Chapter 24

RELATIONSHIPS AND ANALYTICAL FIELD

Madelaine and Willy Baranger proposed the concept of the bipersonal field as early as 1960; it was defined on the basis of Merleau-Ponty's contributions to phenomenology and Kleinian psychoanalysis.

According to Manfredi and Ferro "the concept of field considerably expands that of relation, extending it to the entire analytic situation, that is to setting and to rules, giving a wider view, in the sense that we can consider the events in the "room" before they are taken into the relationship....a sort of intermediate area where scenes and characters come to life and take shape which would otherwise remain locked up and harmless. - From this perspective relationship is a function of the field itself" (Manfredi, Ferro 1990).

The Barangers recognize only the unconscious fantasy of the couple structured via the junction of the projective identities.

Therefore there is a psychic life but also a relational psychic life resulting from the analytic encounter; hence even the areas of resistance are coupled phenomena, blind spots or bastions.

The unconscious coupled fantasy is to be interpreted as something quite different from the single person fantasy. Even though this structure is generated by the instinctual contributions of both the protagonists of an analytic relationship, it does not represent their sum but is rather an emerging reality.

The field gains indeterminacy and evolutive possibilities mostly when the phenomenon of projective counter-identification is limited through the interpretation of the couple fantasy which can occur thanks to the "second look", a specific function of the analyst.

The second look is possible because, while the patient is deeply immersed in the emotional bi-personal field, the analyst maintains a certain capability of analysis which is both self-observation and hetero- observation.

The analyst has therefore the function of contributing to mobilize the dynamics of the field, revealing the distortions connected with projective processes.

The therapeutic effect consists of the possibility of regulating the projective and introjective processes in order to weigh what the patient can re-introject through the interpretation of what has been projectively attributed to the analysts.

Interpretation re-configures the whole bi-personal field and restores to words the original communicative function of favoring discernment, that is, the realistic perception of the analyst's qualities without the distortion of fantasized characteristics.

The psychoanalytic situation is and must remain ambiguous in time and space, thanks to transference shift; it is a common *Gestalt* of unconscious and conscious experiences of the couple. This *Gestalt* must permit the analyst to get caught up in the impetuous whirlpools of the pathology of the field generated by the vicious circles of the patient's life, and to resurface thanks to the analysis of the current processuality.

The dynamics of the psychoanalytic situation is therefore based on projective and introjective dynamics; the analyst and the patient reciprocally get involved in a transformative journey (systemically coupled and autopoietic).

Insight itself is a phenomenon regarding the field and results from the dialogue between patient and analyst; it occurs if there is a common willingness to let it emerge and , according to this view, it specifically consists of a re-labeling of contents, previously split off and placed away from the subject.

It is inevitable or rather necessary that the analytic relation produces a certain level of regression permit that past experiences can be revived.

This regression establishes a narcissistic satisfaction which energetically powers the treatment, contributing to solve the conflicts and re-establish the object relationship.

It is different from pathological regression as a working agreement guarantees a continued value; the patient is certain that the therapist will not permit any evolutive closure through their intervention. Hence trusting the analyst the patient willingly submits to a partial regression for the duration of the session.

Analytic regression entails the emergence of latent contents, according to a logic which follows no linearity or genetic sequence.

This is a salient feature of the relationship: the cotemporality of the genetic phases of psycho-sexual development as they settle within the relational dynamics.

Therefore the field theory perspective questions the sequentiality of the analytic exploration which does not inversely highlight the phases of healthy and pathological development as had been assumed.

Regression is function and result of two regressions: patients and analysts, if it is true that a real analytic knowledge can be reached only if the field is, on the whole, bi-personal.

At this point, some considerations must be made on the temporality of the analytic process in which we revive the non-linear nature of the emotive-cognitive transformations occurring in the patient.

It can be deduced that the transformations are non-linear as their temporal cross section reveal areas of experience which are greatly diversified and non sequential from a temporal (evolutive) point of view.

In logical terms, furthermore, relations between the contents of interpretations and the corresponding mental experiences are not only comparable to causal relations but also to “affinities and involvements”.

It seems that but “here and now” can exist in the analytic relationship, that “here and now” which represents the only temporal breach through which we can work and interact.

The Barangers propose, as a corollary to this hypothesis, that it is more important to experience and process in the here and now of the transference process the memories of the primary scene rather then simply become aware of them; thus the function of analysis does not consist so much of recreating the past but of promoting change.

Past, present and future seem associated with different levels of experiential imbrication and are in the present of the relation "reconfigurable" thanks to the marked ambiguity of the temporal dimensions, the keystone of the evolutive possibilities.

"Past and present discussions to determine whether the analytic process develops and must do so in the "here and now" of the transference state of the session or if it is aimed at recalling memories, seem to disregard the real Freudian dialectic on temporality. If analytic work is possible it is so because the patient and the analyst think that exploration of the past permits opening up to the future; because complementary series do comprise a mechanistic determinism; because, thanks to interpretation, it is possible to escape from the eternal atemporal present of unconscious fantasies. Regressive and progressive moments happen together and condition each other reciprocally (...). Analytic work is played out in the "here and now", in a dialectic between the closed, repetitive temporality of neurosis and fate and the open temporality of insight. (Baranger, 1990 b)

In the Barangers' vision, the more temporality of insight is open, the more it concerns the interpretative recovery of the bastion, the structure resulting from unconscious collusion of the fantasies of the two protagonists of the field, which immobilizes the process.

As soon as "the bastion is the most conspicuous clinical sign of the repetition compulsion" it reminds us the intersubjective nature of analysis" (Baranger 1990 c).

The non-linear temporality which follows from the Barangers' considerations resumes the complex processual structure considered in the Freudian concept of *Nachtraglichkeit,* according to which the pathogenic action of trauma is not to be considered a simple "cause which remains latent until it has the opportunity of arising but a retroactive causation, of the present on the past. (...) The introduction of *Nachtraglichkeit,* indicates those moments when Freud puts aside the model of the mechanistic causality and linear temporality, based on the past-present vector in favor of a dialectic concept of causality and of a "spiral" model of temporality, where future and past condition and give meaning to each other in the structuring of the present" (Baranger 1990 d).

Chapter 25

THE CONCEPT OF FIELD AND COMPLEX LOGIC

The previous considerations directly lead to a new way of observing nature, which not only psychoanalysis but also all modern science uses to understand their objects of study.

It deals with complex reality, which can be studied from infinite points of view and which, for this reason, does not outgrow its own definition but it is enhanced by the contribution of cognitive models which, from time to time science identifies, dissatisfied with what it already knows.

Isabelle Stengers, who, with Ilya Prigogine laid the basis of reasoning on complexity, adds that complexity is not even an epistemology; it is a way of studying problems posed by nature.

In particular she maintains that the notion of complexity does not have an epistemological charter comparable to that of scientific notions in the proper sense. It does not belong to a particular theory or discipline but belongs, rather to "a discourse on science". Therefore, the discovery of complexity in this sense, does not regard the solution of a problem so much as "the awakening to a problem" a "consciousness rising" which has not only an intellectual but also an ethical and aesthetical value.

The notion of complexity also refers to the necessity of a deutero-learning and to the possibility of taking seriously the fact that, not only questions and answers can change but also the kinds of questions and answers through which scientific research develops (Bocchi, Ceruti 1985).

The concept of field is a fundamental question; it is even a pre-complex concept, being used by the Gestalt theory long time ago; here it has a less static acceptation, it refers to dynamics of transformations;

As Francesco Corrao explains: "this new way of thinking also takes into account the observer and their recording systems. According to this new approach, the observer is not supposed isolated; we consider, instead, an observer who endlessly produces a new situation which is to be theoretically described as a new state of the observed *field.*

Every observation is the choice of a particular event (...) *it consists of a desired distinction and in a sense conditioned by how the observer wants to use it and by the context the observer wants to take into account. For example if I study the effects on intersynaptic mechanisms of introducing a certain substance into the nervous system I will be able to prescind from the gravitational forces which still weigh on the system.* Whilst the existence of this distinction is a condition essential to human knowledge, its position remains to a certain extent arbitrary, and is caused by opportunistic reasons and therefore it is partly free".

The physicist Niels Bohr, already in the 1920s further clarified the conventional relation between subject and object of observation in scientific discourse: "The description of our mental activity requires on one hand an objectively given content to be placed in opposition to a perceiving subject ; this is an unsustainable hypothesis as the latter is part of our mental content *exactly like the idea of the object"* (my italics) (Bohr 1927)

This is something new for science arising from the study of atomic phenomena which confirms its validity also in the psychoanalysis field.

These are completely different fields united by the same paradoxes, which render subject and object as comparable abstractions, unavoidable and indispensable to every scientific description.

Even if subject and object are mutually dependent abstractions which, from a descriptive point of view, the behavior of the object cannot always be distinguished from the means of observation, that is, the observing subject.

This highlights a characteristic of the phenomena involved.

There is no point in speaking of specific causes and effects; it seems that since the initial observation phenomena are from the start part of a field where observers are mutual observers and similar generators of further phenomena which include them. They are precisely field phenomena, where observer and observed are an expression of the same emerging event (emerging quality).

In psychoanalysis "every field phenomenon has a bi-personal quality, the patient's pathology, as such, enters the setting only in relation to the analyst who, in turn, contributes to a certain extent in comprising that field pathology which will be the real object of the observation and the analytic elaboration" (Bezoari, Fiamminghi 1995).

Once again physics set out a methodological path when already human sciences were facing descriptive limits with no suitable conceptual tools and without formalizing or being aware that they belonged more to field sciences than non- linear causality.

It can be hypothesized that human sciences, in competition with exact sciences, were determined to defend their positivist and deterministic charter.

Again Bohr in 1938 unequivocally explains: "the elucidation of the paradoxes of atomic physics has disclosed that the unavoidable interaction between objects and measuring instruments sets an absolute limit to the possibility of speaking of a behavior of atomic objects which is independent of the means of observation".

The new way of understanding natural experiences, including psychological phenomena , we reassert, goes beyond the capability of an absolutely objective description "which can only account for the behavior of systems unperturbed by the means of observation" (Bohr 1938).

Later Corrao translates the concept of field by analogy with experimentation in physics, thus defining it "in its wider meaning, the field is a function whose value depends on its position in space-time. Furthermore a system with infinite degrees of freedom possessing an infinite number of possible determinations which it assumes at every point in space and at every instant of time".

The field model does not appear to be delimited by any circumscribing horizon and it doesn't appear to be confined by any perceptive factual observational viewpoint, but it refers to eventual phenomenological movements, invisible in anyhow case deductible and symbolizable through a chosen language (Corrao 1986)

Similarly to what happens in the physical field of phenomena, there are characteristics specific to that particular field in that particular context, characteristics which are preformed according to the functional rules of that field: identifiable rules as in the case of the study of magnetic fields or agreed rules together with "unconscious rules" as in the case of the analytic field, where two real people interact as subjects-objects with all their conscious and unconscious psychic complexity, constituting an inevitably multi-individual interacting structure.

As the Barangers remind us, agreed rules are for example "the initial agreement on the duration of the sessions, their frequency, and the pre-arranged suspensions (holidays etc...) which can interrupt the uniformity of the field. Nevertheless, the analyst and the analysand who start working together know that their work, barring the unexpected, will last some years. Even though they make a commitment for a period of time defined along general lines, innumerable temporal modifications can alter the field; one of the best-known is the phenomenon of the transferentially or counter-transferentially long or short sessions."

Moreover, the field is configured by the systems through which it is structured and develops, such as, in general, interpretative and transformative acts, the passage of projective identifications between the analysand-analyst relationships.

Even at a more basic level, "the essential ambiguity" of the analytic situation , as defined by the Barangers, leaves the field open to any possible evolution, enables the evolution of autopoietic and non-linear processes, typical of the kinetics and transformative dynamics of the field.

This happens thanks to the suspension, of the logical principles of rational thought that is the principles of identity, non-contradiction and casuality, in accord with the fundamental rule.

That is why in Corrao's vision, as I said earlier, the field does not appear to be *circumscribing* but is facilitated by evolutive bonds which allow unconscious contents and a new psychic structuring to emerge.

Therefore ambiguity is a facilitator of processes, a promoter of catastrophic transformations, as Bion explains, referring to Thom's metaphors of topological mathematics.

It can thus be assumed that field phenomena, far from being anarchist or random, are processes free to organize themselves around the bonds which support them by promoting their evolution.

Evolution is constructive in the sense that it allows the emergence of new possibilities for the experiences occurring therein. It is immaterial because its object "no longer consists of deterministic events or products, but atoms and "lexemes "of information and knowledge which reproduce, develop and die like cells of a boundless and complex organism" (Ferri).

In field phenomena scientific explicative models only show maps of possible evolutions of phenomena. Linear determinism gives way to uncertainty, to the uncertainties essential to transformations.

The field develops via possible bifurcations, discontinuities which escape classical mathematics and mechanistic theories specialized in predicting phenomena.

The field develops irreversibly, causes and effects become disproportioned, in short, there is no more linearity; we are in the realm of the possible, of non-linearity.

Therefore the field behaves like a complex system, that is to say in thermodynamic terms, like a system which exchanges energy and information with its surroundings, which is in constant evolution, dynamically unstable and tending towards a growing complexification and where the contingent reality results from a "selection" among numerous possible ways.

Chapter 26

TRANSFORMATIONS IN THE ANALYST'S CONSULTING ROOM

This brief overview of the new epistemological approach to the observation of natural phenomena shows us the path to take to study some psychoanalytical concepts from the perspective of complexity

It is not merely speculation but rather identifying theoretical-clinical tools useful to improving the understanding of phenomena and having a bearing on the effectiveness and efficacy of interventions.

The combination of analytical and complex views leads us to a binocular observation which revises the classical metaphors applied to describe analytical work.

Phenomena occurring in the analyst's consulting room are comparable neither to the archeologist's extractive work – according to the comparison used by Freud in *Civilization and its Discontents* nor to a game of chess according to another metaphor used by Freud in *On Beginning the Treatment* (1913), when analyst and analysand know only the rules of the game.

These metaphors are close to reality, the latter more than the former but in the light of thought on complexity transformations occurring in the analyst's consulting room can be compared to something much more unpredictable and complex.

In analysis, we know only of some bonds which represent the general rules of the game (setting): such as the fundamental rule of abstinence or the frequency of the sessions. However, we know nothing about its development but that it will be evolutive and that the transformations will be non-linear.

Therefore, unconscious fantasies, projective phenomena (specific rules emerging from the analytic game) and all that substantiates and binds the field can evolve unpredictably. The introduction of any variable can make the system function following uncommon rules.

The metaphor of the role-playing game seems, then, more suitable to represent the transformations occurring in the analyst's consulting room and to describe, with a good approximation, the natural development of phenomena. Participants are able to change the powers and possibilities for the actions of their characters and their scenarios at every move, thus continuously redefining the game and its rules.

More specifically, what kind of phenomena and transformations are they? What is transformed? What level of reality are we referring to, these phenomena being non-material and related to the dynamics of emotions and meanings?

The analytic field structures and transforms itself according to "emotional flows", carrying unconscious fantasies of individual and couple which substantiate the field at any moment of its evolution and lead to the emergence of certain verbal and non verbal contents.

The observing function of the therapist proves to be an active and interferential function of the field of projective and spontaneously introjective identifications.

Therefore, analytic observation can, on its own, cause a perturbation, but becomes increasingly mutable as it becomes interpretive, returning the projections in a form comparable to the id.

Scissions are overcome; pathologic crystallizations dissolve precisely through the dialogic therapy which acts recursively on the whole emotional field, facilitating reconfiguration (comprehension) even of the counter-identificative aspects of the therapist.

The whole emotional field is indeed freer to function, that is to symbolize what has yet to be thought and/or to prepare it for a future symbolization.

In other words, the insight which follows interpretation relieves part of the field of symbiosis from the mixture of unconscious projective and introjective functions, reorganizing the structure of the emotional field (Baranger1990). In this way a real symbolic communication is re-established, with new detection of both components, freer to consider themselves and evolve.

An interesting contribution by Cargnelutti and Muratori resumes Popper's concept of propensity: it considers the analytic transformation of a mental state at a certain moment as the probability of revealing itself as a function of the contextual conditions in which analyst and patient are inserted; conditions created by the particular configured setting and by the psychoanalytic theory which has been applied (Cargnelutti, Muratori 1988).

This approach corresponds to the Freudian concept of overdetermination. What that epistemology could not grasp resides in a) the non-combinatory nature of the processual propensity, b) the paradoxical nature of transformative analytical processes.

Regarding a): study on non-linear equations has shown that, even when few variables interact in a system, the predictability of the emergence of a particular state of transformation is next to nil and, theoretically speaking, should the same interaction of the same variables occur, this could produce a different transformation.

If the number of interacting variables were practically infinite, as in the analytic relationship, the propensity would be nil. This indeed means that in this kind of context no combinatorics or predictability is possible.

The more the process evolves the more it should become random; instead the more it evolves the more profoundly it becomes recognizable and knowable. This is the paradox of the analytical process. Therefore certain invariance must necessarily be maintained which permits the transformation to avoid being reduced to a chance meeting of variables and allows certain predictability, even if combinatorics are impossible.

This is an antinomy, a paradox implicit in the overdetermined (multicausal) nature of the process, which does not deny the validity either to propensity or to infinite generativity.

In other words we can say that the more a state of an analytic process alienates from itself evolving through interpretative transformations, the more it realizes itself.

That being a complex transformation, a further phenomenon might take place: a massive, congruent and simultaneous transmission of change to every element substantiating the field, similarly to what happens with other complex studied systems for example, biochemistry.

Systems sometimes transform themselves according to the general conditions of the entire field in which they organize themselves.

An example can be found in the experiment of the chemical clock described some years ago by a group of researchers at Brussels University.

In this experiment the concentrations of two substances reacting under specific chemical conditions imposed on the entire system, instead of behaving chaotically behave as a whole, with coherent changes at regular intervals.

It is as if molecules changed their properties simultaneously; they behave as a whole, seemingly to communicate at a distance. (Prigogine 1993).

This is a quite singular phenomenon contradicting all knowledge about the functioning of chemical systems; it is actually a property of the non-linear transformations of complex systems.

To conclude, the *generativity* of the field is recursive (autopoietic) that is, causes and effects are also at the same time causes and producers of what produces them (the field). This is an explicative *quantum* beyond the explicative possibilities of cybernetics and of the theory of systems which seems specific to living organisms. The resulting generative transformations are non-linear, they occur with a disproportion between causes and effects and they produce new equilibria through unpredictable bifurcations and multi-furcations.

The field devours disorder in order to create thinkability organization; the new structure can contain areas of disorder which are symbiotic, not thought out and which do not destroy the system but continuously nourish it.

The whole pervades the parts in order that change and stability can co-exist. Nevertheless under particular conditions of instability an even infinitesimal part can affect the whole leading to a general reorganization of the macroscopic functions.

Finally, generative transformations can function communicating homogeneously at a distance in the various areas of the field itself, inducing massive behavior consistent with the entire phenomenal domain, as we will see soon, touching on the vicissitudes of group field theory.

Projective identification is an example of massive, congruent and simultaneous transformation in the analytic field.

The emotions the actors of the field share through projection are similar, induced and maintained through a sort of projective reciprocity, where emotions exchanged at an unconscious level, tend to be self-maintaining.

A feeling of unconscious hate, for example, can generate a projective counter identification of hate which can show through non verbal behavior; this exchange will inevitably return to sender as unconscious confirmation of the projective hate.

Chapter 27

LEARNING

The mind, when faced with sensory interoceptive and exteroceptive impressions, is in a state of anticipation or *preconception*; the combination of suitable sensory impressions and preconception gives rise to *conception*, on which the successive development of thought is based.

According to Bion, repeated combination of sensory impressions and preconceptions generates an apparatus supporting both knowledge and the capability of continuous learning. It originates from the internalization, of projective identification governing the mother-infant or patient-analyst relationship.

Since projective identification always comprises a mobilization of emotions going through the emotional field, involving both the protagonists of the relationship, it is deduced that learning and development of thought are ruled by interaction between affective and cognitive factors.

Thanks to the fact that preconception remains insaturable, it can actually constitute the basis for further learning, achieving what Bion defines as the apparatus for thinking.

It functions via the flow of further emotions which Bion indicates by "+" and via the ability to tolerate doubts, that is to the lingering of the apparatus in a state of expectation, indicated by ".".

"This is the fundamental psychic condition for the individual who can maintain their knowledge and experience, and at the same time be able to rebuild their previous experience in such a way as to be able to conceive new ideas".

Thus learning complexifies, even though the receptive qualities of preconception remain unaltered, thus a constant conjunction of container and contained is achieved, necessary for learning from experience. (Bion 1962)

Every sensorial perturbation permitting development of concepts is subject to the maintenance of the organization and the development of the apparatus for thinking, at least under conditions of psychic normality.

The continuous production by the apparatus for thinking maintains identity and invariance. Essentially, the process which organizes thought self-reproduces evolving and by permitting in its own organization only those perturbations which allow an increase of knowledge.

In other words we can state that the organization of thought generates more of same and of further components of which it comprises, generating with no predefined limits and towards an unpredictable development.

Ultimately, also the process of learning from experience follows autopoietic logic.

Chapter 28

THE FIRST STEPS OF THOUGHT

> If there were a means to choke the fount of hearing through my ears, I would not have stayed my hand from locking up my miserable carcass, seeing and hearing nothing. It is sweet to keep our thoughts out of the range of hurt.
>
> Sophocles, Oedipus the King

According to Bion , at the proto-mental level, turbulent phenomena continuously occur, phantasmic vortices of β elements requiring α functions to be transformed into images and then into symbols.

β elements are course sensorial information pre-existing thought, even though they simultaneously comprise its prerequisite which requires an organization.

It is interesting how in some psychotherapies we can identify the precise context where the proto-mental level of thought emerges and how, at the same time, the sensorial order combines with a primitive and basic body-cognitive experience of self.

The proto-mental level of thought mostly emerges during the treatment of psychoses and borderline syndromes, thanks to emotional participation, that is, clinical empathy.

The treatment highlights that these pathologies particularly endanger the self. Its sense of uniqueness and distinction is constantly on the verge of disintegration, destruction (Lichtemberg, Lachmann 1992).

This is, first of all, a matter of disintegration of the sense of physical unity, in the sense of sensorial perception of self.

If Ego is considered above all a body Ego, this can also be interpreted in an ontogenetic (temporal) sense; therefore the representation of self is above all physical, that is mediated by basic sensorial representations: visual, auditory and kinesthetic.

If these sensory representations engage the α function they also provide independent variables which recursively contribute to the formation of α elements, that is to the construction of more organized levels of experience.

In substance, "the various psychophysical levels equally contribute to the complex organization of the organism. Rationality, memories, the symbolic world, postures and movements, the emotional world and finally physiological systems are all deeply integrated and interconnected psychophysical functions."

Mind-body function does not seem to be pyramidal "with the mind controlling everything from the top, but recursive"; this perspective, too, is complex: observation points are reversible, the psychosomatic unity is autopoietic: body perceptions and sensorial experiences modify higher level representations, which, in turn, modify the perceptive level (Rispoli 1998).

In a recent work Antonello Correale argues that "empathy is the capacity to continuously tune in to the patient's wounded and damaged Ego and to the means they adopt in order to adapt to the reality and to defend themselves from this condition. In other words it is useful to interpret as empathy the capacity to grasp the patient's deep need for vital narcissistic gratification" (Correale 1995).

In my clinical experience, I have had the opportunity of noticing that with seriously ill patients empathy needs organizers of thought which can be provided by the therapist in the shape of metaphoric images.

It is also necessary though, to help the patient gain and keep a sensorial representation of self, in order to contribute to containing the body ego dissolution, which tends to erase the background, the basis on which the metaphorizing organizer of the apparatus for thinking can be established.

I think that this preliminary work can be useful also for ambulatory psychotic patients and those borderline organizations which efficiently hide the psychotic nucleus.

As an example I will mention a short clinical episode concerning a patient with a good social and adaptive functioning but with a seriously damaged sense of identity, hypochondriacal symptoms, bulimic and anorexic behaviors, various problems with the modulation of aggressive drives.

Clinical Report

Alice has been in therapy for three years. At the beginning of the session she tells me about a film entitled "*At First Sight*". A long series of associations develops about the importance of being able to see; she reveals that she has managed, thanks to the analytic work, to see herself no longer through her mother's or her boyfriend's eyes but through her own eyes, despite the fact that this is very painful.

Fulfilling every expectation (cognitive-affective domain) of the person observing her (perceptive domain) has always given her the certainty of existing.

During the sessions, which take place face to face, seeing has always had a central function. Alice thus tries to resume her search for perceptive-sensory coherence by grasping a confirmation.

While she is associating, I remember some considerations she made some time ago based on a book I read: "*The Girl Who lost her Voice*" On that occasion she explained that words also had the power to make her exist.

Coordination of visual and auditory representations of self is the basis of a growing proto-cognitive domain.

It seems that, by trying to grasp the experience of self at these perceptive- sensory levels, she indicates the basis of the psychic contents I must also get to permit her to feel real, cohesive and verifiable through exteroceptive perception.

She seems to say: I am starting to exist because what I observe is myself, because I speak and I listen to myself speaking, and this is all unquestionable like you, in front of me, are.

It is the beginning, a primitive *self-cognitive achievement from* which cognitive-affective representation of self can develop.

I point all that out. Alice cries and compares me to an angel dispensing sweetness.

I think she experiences me as a hypostatized and asexual entity, an archaic container which owes its salvific function to being an aggregator of the developing self.

I think that the primitive idealization invested in me is an evolutive request, meaning that the object-self accepts the needs for mirroring and idealization to be the basis of a coherent self (Bach 1985).

It seems interesting to underline that in the reported case very different levels of perceptions, cognitions and emotions have contributed to a primitive structuring of a coherent self; some of them are quite evolved and have been for a long time at the service of the principle of reality and of the symbolic regime, for example the capability of verbally self-describing the structuring which is taking place. Others are just emerging from the proto-mental field, such as the use of the perceptive function aimed at a visual and auditory representation of the self.

The self-recursive process which has structured the self uses mind-body functions belonging to both the actors of the psychoanalytic relationship and to different levels of deeply interconnected functional complexity.

In conclusion, the autopoietic organization of the field does not establish itself through an accumulative sequentiality of increasingly complex structures but through distributed and heterarchic sequentiality.

Chapter 29

PROJECTIONS

If we dealt with a numerical system we would speak of units of measurement speaking about emotional and symbolic concepts we can find a unit of analysis of the phenomena of the field, by analogy, in the projection.

Using the indefinite article I want to underline that we could use other equally pertinent observation points such as cognitions and their distortions, insights, defensive structures etc...

This said, the intersection, combination of mutual projections, and the couple fantasies substantiate the analytic field and power its dynamics.

Let us focus on what I mean by projection and at what level it can be found if observed from the perspective of complexity.

According to Freud it is a very archaic mechanism, being present in an individual from birth and used as a defense from stimuli from which they cannot otherwise defend themselves

It does not exclusively have a negative connotation; on the contrary it plays an essential role in the development of personality as it accompanies its reciprocal, the mechanism of introjection, through which Ego develops and assimilates the qualities of primary objects.

Similarly to what happens in instinctual life, projection-introjection permits the rejection-introjection of experiences as if they were a child's unliked food.

Freud, did not completely solve the question referring to *what* is projected; as Laplanche and Pontalis remind us , it is the feeling of hate, experienced as if it were coming from another object; therefore, as we can see in "*The Clinical Case of President Schreber*" the statement "I hate him" is transformed into "He hates me".

In works such as *Drives and their Fates* (1915) and *Negation (*1925*)* it is the hated person who is projected externally; therefore the acceptation already refers mostly to the object or to the relationship with the object rather than to the drive (Freud 1915, 1925).

In Melanie Klein the process develops: if some content is to be refused it is necessarily connected to an object rejected through projection (Laplanche, Pontalis 1967).

In particular, the complexification of the projective process is completed through the concept of projective identification.

According to the original meaning of projective identification (1946) "The subject totally or partially introduces themselves into the object in order to damage, possess or control them".

As far as projective identification is concerned, we cannot prescind from the presence of another person; it is a phenomenon always involving two people. Through this process, to avoid total loss, what the mind itself rejects while remaining in touch is transferred to another, creating a pseudo bereavement.

Moreover, projective identification implies the empathy of the object. It urges the object to fulfill the projective request, thus favoring the introjective side of the identification process (Grinberg 1975).

With Bion the concept is further developed: not only the structures of the personality are projected, but also entire functions of the Ego are split and projected: "thought, memory, attention, verbalization, action, judgment". He thus studied the self-destructive function of the mind which indeed "can attack itself, fragmenting minutely" (Meltzer 1993).

Moreover with Bion for the first time the transformative relational aspect related to projective phenomena is described in a well-structured way. Bion thinks that the mother is the first transformer of the unbearable thoughts in the child's mind; the mind thus augments itself somewhat as if the structure of thought from now can extend itself to a matter of two individuals, exactly as happens in the analyst's consulting room, where thoughts are offered to the transformation by the analyst's mind, which *predigests* those unbearable thoughts.

The field of thinkability thus develops and, more importantly, the function of autonomous thought is introjected, that is an apparatus for thinking is created via the introjection of another apparatus already fit for the purpose.

The mother, as also Meltzer reminds us, is not simply a *providing* object but she is fully a *thinking* object, who participates via her thoughts in the construction of those of the infant' thanks to what the latter will be able to dream and symbolize. Bion defines this, the alpha function.

With Bion we move into an area of thinkability which deals with relationships, we take a step forward towards that conception, exhaustively developed by the Barangers, wherein the field and no longer the relation is the cardinal concept. From this moment on projective couple phenomena comprising fantasies will be conceivable.

Bion is important not only because he points us in this direction, but also increases the possibility that projective thoughts can be considered as group phenomena; with Bion there is no longer an individual or a couple limit to the fantasies underlying the projective phenomena.

The dimension is *plural* now, a group being a totality itself.

This can be fully observed in *Experiences in Groups,* where the mentality, the culture, the temporary structure, the organization and the tasks of the group are simultaneously the structure, mentality and organization of the individual.

The mentality of the group is structured on the so-called basic assumptions among which Bion explains three: the assumption of dependence, of fight/flight, of pairing. They represent proto-mental levels of psychic activities as they belong to both the individual and group mind which has not become emancipated yet from the undifferentiated matrix from which both psychic and physical phenomena derive.

Here we are on the border between the mind and the body field, which Freud has already shown us; it will not surprise us to be facing group splitting and projective phenomena, which are clearly observable when studying the scale level of social psychology.

In clinics, too, when the level of analysis is extended to work groups we meet similar processes: the split and projected parts of a subject generate group dynamics which follow the same splits, acting as a sounding-board of individual splits. These dynamics, indeed, if not metabolized by thought, backfire on the subjective nucleus, amplifying intrapsychic malaise.

Moreover it is common, especially in the treatment of serious pathologies that these phenomena travel like a wave, invading the whole group field and creating real therapeutic blocks.

Therefore projection is not an isolated phenomenon, but a peculiarity of the field; it does not refer to an object considered as an individual, but it is specific to an "intrasubjective object"; something which slowly takes form, resulting from the development of multiple relationships involved in its structuring and which only in the extreme can refer to a possible "primal scene" and to a *primal intrasubjectivity* (Gesuè, Sabucco 2000).

Projections have so strong an effect that they change the involved minds, they are, indeed, the most important means of change, the means par excellence with which the analyst has to work during a session.

By working through *containment* the analyst carries out a radical transformation which, according to Ogden, enables them to act even in silence.

Ogden specifically reminds us that "It may be that the essence of what is therapeutic for the patient lies in the therapist' s ability to receive the patient's projections, utilize facets of his own more mature organized personality system to process the projection and then make the digested projection available for reinternalization"

This containment does not always occurs; the effectiveness of analytic transformations lies indeed in making projected contents conscious, that is thinkable; otherwise nothing changes in the field: everything perpetuates according to an endless repetition compulsion.

The receiver may not be able to live those feelings and does not metabolize them. They don't verbally expressed dealing with them, instead, through denial, projection or projective counter-identification.

"In this case the projector would be convinced in his belief that his feelings were indeed dangerous and unbearable. Through identification with the recipient's pathological handling of the feelings involved, the original pathology of the projector would be further consolidated or expanded".

Where thought does not succeed in gaining position, projective identification agglutinates the experiences and prevent change because "in projective identification, not only does the patient view the therapist in a distorted way that is determined by the patient's past object relations; in addition pressure is exerted on the therapist to experience himself in a way that is congruent with the patient's unconscious fantasy" (Ogden 1991).

When the analyst is victim of projective identification they retroact in turn, as Grinberg would say, through a projective counter-identification, unconsciously interfering and maintaining the *agglutination* of experiences, fantasies and projections.

Therefore the concept of projective identification proves to be something quite concrete, nearly a *concept-action-thing.*

Due to the fact that projective identification is a concept-action, it differs from the concept of projection as developed by Freud.

In projective identification there is something related to doing, expelling in order to cause a change in the destine of the projection and to recover a modified part of the projection.

It is therefore an intersubjective concept, concerning the field; it cannot exist in vacuum, but only in the presence together of at least two minds in a relationship.

What is done in the relationship is done concretely, pushed, induced, forced and, as we have learned from field dynamics at times co-created in the other and with the other; it has effects even on the respective mental apparatuses.

This projective mind modifies itself, it removes the part put into the other or delegates to the field to again get involved, positively if there has been a good containment, negatively if the resulting introjection has not been metabolized by the other's mind.

We must remember that, apart from a clear defensive function, based on primitive splitting, projective identification also has a communicative value, as a signal useful to therapeutic processing leading to a healthy reintrojection (Kernberg 1976), (Langs 1976).

In Bion and Ogden projective identification acquires a connotation of external communication-action between distinct mental apparatuses; in Meltzer as an evolution of thought based on Klein's unconscious fantasy, the concept assumes a complementary meaning, becoming a tool for clinical observation of the intrapsychic.

On the basis of this assumption I will claim that projective identification is a very complex intra-inter-co–subjective concept, comprising a range of multiple and simultaneous dimensions.

Chapter 30

INTERNAL PROJECTIONS

Donald Meltzer explores a new dimension of the projective identification; he studies this phenomenon by working especially with children, in a field where phenomenology, hidden behind concepts, definitely becomes concrete and accessible, revealing the meaning behind the words.

The way the infant uses the analyst and the objects in the consulting room indeed highlights the unconscious fantasy underlying the projective act, differently from any other patient typology.

By working with children, as Meltzer reminds us, one realizes how, through their projections, they can find a way into the analyst's body.

It can be noticed through the way the infant rolls around on the floor or contacts the analyst's body or how they observe the analyst, trapping them or trying to get into them.

In these cases, one is aware of how intrusive identification acts, aiming at establishing the base for the contents which are projected onto the inner organs of the object.

It can be hypothesized that the interior of the object is compartmentalized, divided into sectors, each having its own psychodynamic value.

In *Sexual States of Mind* Meltzer introduces compartmentalization of the inner mother founded on the body fantasies connected with body orifices (Meltzer 1973).

In *Claustrum* and in *Working with Meltzer* the theory takes a definite shape in a study on claustrophobic phenomena (Meltzer 1992, Meltzer 1998).

Meltzer here has the implicit ambition to substitute classical, purely descriptive nosography of sexual pathology with a decidedly metapsychological one.

This systematization explains the primordial psychic levels, according to a *spatial topology* based on unconscious sexual fantasies referring to the various parts or sections of the referent parental body in the form of partial objects connecting, mating, or combining to construct a fundamentally bisexual personal adult identity.

In serious pathological cases progress can be achieved only through a deep dependence on the analytical breast; only this way can the psyche reach a depressive position and object constancy.

From a different point of view we can say that only projective internal identification with the object-breast is able to modify the representation of split unconscious fantasy.

The capability of projecting unconscious fantasies related to zonal compartmentalization of vitalizing objects is extremely important to intra- projection: they give back evolutive possibilities to inner, devitalized sclerotic, affectless objects.

In paranoid anxiety, in particular, the persecutory element of the internal psychic component split paralyses the vitalizing function of the good breast which counters the evolutive opportunities above mentioned.

In substance it is the omniscient reassuring illusion of the destructive component which develops as an addiction elicited by the impossibility of coping with separation.

The fear of this interiorized persecutory psychic component is reinforced by the internal projection of the bad breast paradoxically becoming itself a protector from the fear of loss.

Better the fear of persecution by the bad breast than the fear of its abandonment.

This happens in perverse sexuality, in a narcissistic organization which is difficult to dismantle.

This assumption is based on the fact that the internal mother, initially undifferentiated, progressively compartmentalizes, depending on the experiences the infant has with their orifices according to how the mother treats them during daily body-care.

These various experiences are followed by articulate and complex fantasies corresponding to the involved regions which lead to a sort of internal geography of the object's body, geography or compartmentalization which, according to Meltzer, is increasingly complex the more it is built on hypothesis far from the direct experience of the infant, above all with regard to the internal part of the object.

The process proves to be even more complicated due to the inevitable shortcomings of the mother in fulfilling her caregiving tasks and not least due to the ambivalent feelings and emotions which the exploring of the object implies, with consequent zonal confusions or overlaps of compartmentalizations.

For example buttocks can be assimilated to the breast or the inner parts can be fantasized or confused.

Head, breast and anus can thus create an internal geography, object of the projective identifications of the developing subject and can give birth to projective confusions associated with psychopathology or, on the contrary, they can lend themselves to a form leading to psychic health and balance.

The following is given as example: it is a clinical report on a case I followed in the second year of my university studies in Psychotherapy.

Clinical Report

Alessio is forty-five, homosexual and HIV-positive. He is an imposing man with a body carefully kept fit through gym workout.

He speaks quickly and often flounders looking for a plausible expression, which does not fail to reveal the meaning of the experiences he recounts, experiences he has lived to his amazement, as if events always happened according to a strange and unpredictable logic.

Even though he has travelled around the world and frequently changed surroundings he feels he has never found his place. He often feels confused, aggressive, and suspicious with people and he has sleep problems. He would like to be able to attain a consistency he defines "ontological involving, as we will clarify together, cohesion and continuity in his experiences which would make him feel complete.

After about a year of analytic therapy he tells this dream: I go to take some money under a vase of flowers. The banknotes are wrapped in a sheet of plastic so that they do not get dirty, I feel uneasy...after a while I find myself in a room with a woman, I start defecating and I feel uneasy again.....then the scene changes and I see myself in a square with two churches while I meet Luigi, a childhood friend with whom I have a talk....now there is the steward of the institute where I used to work and I get angry because he wants to keep some of my money..."

The evidence of anal contents in these oneiric images and the work we have done till now justify direct interpretation, hence I ask dubitatively whether he remembers that he had reciprocal anal masturbation with Luigi: Alessio immediately admits that it could have happened and makes a series of associations about his own passivity during intercourse with numerous friends during his childhood.

I underline his manifest emotional passivity currently towards people in his life. Alessio often feels submissive and angry for that, at work he has always had a difficult relationship with his bosses, he secretively attacks them even if he is afraid of their opinion especially as a moral judgment.

After my words he recognizes that many of his personal choices are related to passivity and to his conflictual disposition to be a receptacle of all that others want to introduce, he acts as a sort of cloaca for the bad things others do not tolerate.

Alessio has projectively identified himself with his own partial internal object, specifically with the rectal bin, with that claustral zone masterfully described by Meltzer, and has lived in terror of being *ejected*, that is rejected into a dreadful and unmentionable outside.

I do not intervene, waiting for Alessio to let his thoughts take shape; after a while my patience is paid off.

"I behave like my mother, she taught me passivity, and she was that way with everybody, particularly with my father".

Through this insight he lays the bases of a more conscious behavior, inclined to a more defined object distinction, which will interestingly develop in successive sessions of therapy.

What is important does not consist, as far as we are concerned, in psychopathologic conclusions derivable from compartmentalization of the internal object, to which I refer the interested reader, but instead the fact that, as these studies also highlight, projective identification is a process expanding in various directions, with different meanings (communicative, defensive etc...) and with a connective function for the development of a unified theory of projection as it is a broad-spectrum process, acting at different levels of complexification of internal or external, entire or partial, objects.

In other words, projective phenomena can be found at different levels of observation: from the microcosm of the intrapsychic acting on partial objects, to the cosmos of object relationships with the external object, and then to the macrocosm of the relationships between groups and finally in elements of the field like couple fantasies (which after the Barangers, cannot be considered isolated phenomena but the result of crossed identifications which act and constitute the field).

The combination of projective processes at the different levels of scale constitutes an organized whole, where the constitutive relationships are the same projections; a set with its own structural peculiarities of stability and evolution, which are different according to their context, and which therefore have, from time to time, unique and consistent bounds and possibilities.

It represents also a system of knowledge, an active organizer of environmental data (internal and external objects) which can be affected and to a certain extent reorganized by new interactions (Ceruti 1989).

We can state that a system of projections is a constructive and cognitive system, as knowledge derived from it "is connected with an action which modifies the object and which does not reach it other than through the transformations caused by this action. In this sense the subject (...) immerses in the object and reacts on the object, enriching it; now subject and object are placed at exactly the same level (...) In short, there is no more distinction in principle between the subject and object; the subject extends itself in its tools, instruments and implements inserted into the object" (Piaget 1967).

Can projective phenomena, via those elements, those building blocks, be the means by which we can study all that is considered psychic? Can these intuitions be the basis of a clinical action taking into consideration the multilevel structuring of the projective organization? What kind of organization is it, comprising multiple possible dimensions?

Is projective organization, in the light of modern psychoanalysis, merely multilevel or can it adhere to anything different, such as an organizational quality, difficult to define through common topographical categories? We, indeed, discussing the internal and external psychic world, apply a topological metaphor based on Euclidean geometry. Which metaphor best describes the projective dimension and can therefore become an efficient analytical and clinical tool? Is it possible to detect such a metaphor, using words to describe such complex processes, at the limits of thinkability?

Chapter 31

HOLOPROJECTIONS

There are temporary answers, more refined than those previous.
Instants when things simplify, crystalize.
Still the last word is never pronounced.

Alain Connes

According to Edgar Morin complexity is based on three main principles.

The first principle is dialogic; it affirms the co-presence of opposites in systems.

As I have argued in the chapter *The Emergence of Order,* two opposites, order and disorder, coexist to fuel each other in an endless loop whose starting point can be only arbitrarily defined. Opposites coexist, this is the dialogic principle, and remain in an unstable but creative equilibrium, evolutive, unpredictable and tending towards greater complexification of the system itself.

The second principle has been introduced in the chapter *Autopoiesis,* focusing on the recursive process.

According to Morin "a recursive process is one where products and effects are simultaneously causes and producers of what produces them", he gives the example of society, according to which every individual is a product of society and society, in its turn, is a product of the individual.

The third principle is hologrammic; it assumes that not only the part is in the whole, but also the whole is in the part, with no interruption, going beyond both the holistic and the reductionist concept of organization.

Complex systems such as the analytical field exactly follow this principle: if we consider projective mechanisms as basic elements, the hologrammic principle is evident.

Projective mechanisms are applied to the intrapsychic domain, comprising the "connective tissue" of the analytical field; therefore the part, (the projection) is in the whole (the field) and the entire emotional field can be reorganized by a projection made in any part of the field itself.

The projection is thus represented by a hologram.

Ultimately, if the projective process is the unit of analysis of the field, then the organization is hologrammic, recursive and dialogic.

This goes completely beyond linear logic, introducing us to a discussion where knowledge of the part and of the whole fertilizes both, in an absolute relation of complementarity.

If that is true, we must be aware of the fact that every projection reaching the analytical field recursively rebounds on the whole emotional field and, vice versa , all that happens in the field must be expected to be found in the unconscious fantasies and in the individual projections.

There is nothing left for us but an extremely important choice from a theoretical , epidemiological and clinical point of view: to apply Gödel's theorem, according to which no system can explain itself by its own (if systems are complete, wherein they can express every statement (including about themselves), they must contain some false statements; instead, incomplete systems can, by definition, contain only true statements but cannot express all statements (especially about themselves)) and to accept the use of meta-points of view which explain from continuously different vertexes our boundless object, that is the analytical field , with regard to that portion and to the extent pertinent to the context we are observing and on which we are acting through our work.

Chapter 32

BION'S SIGNS AS OPERATORS OF A COMPLEX LANGUAGE

Watchman, what is left of the night?

Isaiah, 21, 11

The sign "2" represents an entity available for mathematical transformations which obtain other derived entities; for example "2" can belong to a set of numbers subject to function transformations.

A mathematical sign such as"2" is always arbitrary and meaningless; even when it is used for carrying a message, this happens by virtue of it being used as an element in a statistical application, as explained by information cybernetics according to which a message is a derivative of the frequency of the emitted sounds.

A message depletes its contents in the mathematical formulation which describes it; it is nothing but a process of transit from a state of maximum disorder or ignorance to a state of organization or knowledge (Shannon e Weaver 1959).

Suppose, for example, the final message ADA is to be codified, which in the common acceptation of the word is a name for a woman.

In mathematical terms the name ADA refers to the result of an informative process given by the difference between the degree of inbound and outbound uncertainty of the memory system or the system of transformation under consideration.

The initial uncertainty will be given by the number 9261, and by convention by its logarithm; in the Italian alphabet there are 21 letters, the possible combinations of the three letters composing the name ADA are $(N)^p$ where N represents the number of the elements of the set and the exponent p is the number of elements in each group of three elements Thus $(N)^{p \text{ equals}}$ 21 raised to the third power, that is, 9621 and Log K = 9621=21. This is also described as entropy (Beccaria 1994).

Generally speaking, the decrease of outbound uncertainty of the system is given by (Log K-Log H), which is the difference of the logarithmic functions of the degree of uncertainty at the beginning and at the end of the informative process.

Consider now the sign ♀ which in Bion's theory stands for "container"; this sign can describe every action, word, gesture, group and infinite other entities which at different moments and different levels of complexity deal with receiving and transforming projected emotional contents.

Therefore the sign, standing for multiple possible realizations, is open to a wide range of possibilities and shows a certain degree of extensibility.

Bion's sign can be used as a symbol, similarly to the cross or the dove, and possesses the property of representing something which means something else, while maintaining a perceptive, formal, allegorical or metaphorical connection with what they suggest.

It cannot be used operationally in other words, it cannot be compared to the sign “2” or to the name ADA, because it cannot easily be defined by a statistical operation since operationality is incompatible with extensibility.

On the contrary, a mathematical sign can easily be manipulated via transformations; but is conventional, arbitrary; it is similar to Saussure’s “independent symbol” with no relation to what it represents.

At a deeper level of analysis sign and symbol cannot be completely distinguished; they indeed have a common etymological root related to the concepts of body and life.

Their common root *sema* refers to *soma,* which means body.

The root of *soma, twere,* generates “to cover” and the meaning of *sarx*: of the body in the sense of flesh. Flesh, in Greek is, *Kreas*, derived from *ker* which means “to cut”. However, “to cut” also originates from *sek* and from it is derived the Greek, *sekos*, meaning consecrated space, sanctuary.

The body is therefore the place of the soul that is the place of the symbol; thus the religious dimension of the body is accomplished. This is further affirmed once we consider the Indian linguistic derivation referring to “festivity”, a festive meeting in honor of a divinity.

Therefore sign and symbol originate from the body. They differ in their depiction of it although always referring to it (Tibaldi 1983).

A symbol is an archaic product, which at the same time distinguishes itself from the body and represents it.

The first communicative wall inscriptions were actually ideograms, symbols whose connection with what they represent lies in their shape, specifically in the body shape.

The symbol differs, then, from the perceptive datum, arriving at an ever higher level of abstraction.

Via the symbol, a high degree of intersubjective validation and subjective validity can be achieved. Using symbols, the subject can indeed efficiently represent and communicate their internal states.

If we imagine placing symbol and sign on a continuum we find the former tends toward maximum fidelity to the percept, to form and not easily rendered operational, and the latter in the direction of total perceptual arbitrariness but highly operationalizable.

Taking this abstraction further, following Bion, we can arrive at the sign

♀ < - > ♂

Representing the dynamic interaction between container and contained, the mechanism of projective identification which maintains both operationality distinguishing the mathematical sign and extensibility of the symbol.

By maintaining the characteristics of both the sign and the symbol a new entity is achieved; it can be used for complex abstractions able to describe the progressive structuring of the bipersonal interactions.

A mere mathematical sign would not account for the metaphorical nature of thought and a symbol would not permit the operationality requested by the description of the interactive process.

The characteristic of the sign

♀ < - > ♂

cannot be reduced to a statistical operation but it is useful to represent the recursive operation of the interactive process which progressively transforms the rough stimuli of the body experience into thoughts.

As can be deduced from the etymological analysis as well as semiology this confirms the "body" genesis of symbol and sign.

Chapter 33

More on Hologrammic, Recursive and Dialogic Organization of the Field

In order to fully understand the hologrammic, recursive and dialogic nature of the emotional field it is necessary to refer again to projective identification and the correlated psychological processes.

The projective process is not only an evacuative defense to relieve the intolerable tensions of one's psychic life but it also contributes further psychic components.

As Ogden reminds us, they who project, communicate, try to explain their own internal world through an intrusive unconscious fantasy which puts real pressure on the target-object in order to make it conform to the projected fantasy which carries a message transformable into symbolic (linguistic) terms, via an interpretative restitution or a symbolic action.

Projective identification is, therefore, a part of a path towards psychological change; by which feelings similar to those the person projecting is fighting, are processed by their receiver.

The assimilation of the projected contents is thus possible, leading to integration, the object of the analytic change.

Projective identification is therefore an *autopoietic* process which positions itself in the relationship between subject and object of the projection where "interactions and transformations regenerate and realize the network of processes (relationships) which produced them (*recursion*); this builds up as a concrete unit (...) as a domain (...) of relationships specified with respect to the autopoietic organization which these relationships make" (Maturana, Varela 1980).

The configured domain is communicative, linguistic and cognitive; it distinguishes itself as a process at all these different levels of organization and assures unity, that is, distinguishability and specificity of its own determinations, at every level and for all levels (as a system).

In other words, the levels of organization are both distinct and unitary, *imbricated*, one cannot exist without the other and interacting with a level necessarily implies conditioning the others.

Even though Bion did not use these words, he realized that the characteristics of projective identification suited neither linear nor circular logic.

Even circular logic seems indeed inadequate to grasp the complexity of projective identification; it detects incoming data and outgoing processed information in a system, but it does not account for the simultaneity of projective and introjective processes and above all it does not grasp their recursive and autopoietic nature.

Bion indicates the relationship between container (analyst's affection and knowledge) and contained (need for knowledge) with the following pattern of dynamic interaction

♀ < - > ♂

To specify its reiteration and generativity he adds the exponent **n**, from which follows

♀^{n} ♂^{n}

which explains how a growing conjunction of container and contained is necessary for the learning experience.

So, the growing conjunction comprises the whole cognitive process: from the minimal experience of knowledge of a basic need to “whole systems of hypotheses known as scientific deductive systems. These extremely complex systems, though their origins are impossible to recognize, retain the receptive qualities denoted by: ♀” (Bion 1962).

Therefore also in Bion's vision projective identification is a complex process, since it combines several descriptive levels and it constitutes the base of more and more abstract inclusions.

He therefore uses an abstraction which “facilitates correlation via comparison between the abstracted statement and a certain number of different realizations, from none of which that realization was initially abstracted” (Bion 1962).

The evolution of learning goes with the development of the apparatus for thinking, that apparatus able to fulfill the α function, which translates the rough data of sensory experience (β elements) into thoughts (α elements); this evolution is also described by $\beta \rightarrow \alpha$.

According to the theory of the field a projective identification represents an *emotional flow, involving* patient and therapist in a vortex of non-thinking which sacrifices the evolution of learning of the couple’s experiences and their respective contributions to it.

Interpreting a projective identification means then, bringing back the couple’s confusional acts placed at the symbolic level at a point inside the field, to the whole emotional field, through a holo-logic process.

The meaning bearing restructuring process concerns, therefore, a part (the analyst’s interpretation) which retroacts on the whole (field) which, in turn, retroacts on the parts (analyst and patient), giving back meanings and new possibilities of evolving (hologrammic organization).

This process is dialogic too, since it envisages the creative co-presence of opposites.

The restructuring, mentioned above, leads not only to the emergence of sense but also of new ground not yet exploited for thinkability yet.

These considerations which enhance present knowledge about complexity are considered the basis of the possibility of knowledge by Bion who, arguing the infiniteness of the constant conjunction, states: "A $♂^{n}$" must therefore have a phenomenological counterpart represented by the concept of infinity.

The elements of the many scientific deductive systems must be capable of recombination. A well-known example of recombination is the use of an hypothesis from one deductive system as premise in another deductive system According to the theory put forward here the freedom necessary for these recombinations depends on emotions suffusing the psyche because these emotions are the connective in which the elements of ♂ are embedded. The ability to tolerate doubt and the sense of infinity constitute in $♂^{n}$ the essential connective which renders K possible (knowledge) (Bion 1962).

Chapter 34

THE ANALYTIC FIELD AND NONLINEARITY

Bion first argued that analytical change activates transformations of emotional experience and psychic structure itself.

Interpretation is the most important means for transformation of emotional experience through contact with the patient.

The analyst has indeed the specific task of transforming latent emotional contents in communicable meanings.

In field theory transformations are a crucial aspect of the new psychoanalysis, as the stress put on intersubjectivity also underlines the transformability of the experience and consequently of the psychic structure.

Weak interpretation is no doubt the most innovative tool of this research; it is a procedure which neither censors nor peremptorily decodes, opening to new intersubjectively validated developments of meaning.

Weak interpretation "does not explicatively convey a meaning already clearly configured but rather contains a highly incomplete semantic planning, which can be realized and better defined only with the patient's active contribution." (...) Moreover "interpretations gain shape and meaning thanks to an *hermeneutic cooperation,* wherein the patient is in a relationship with equal - although not symmetrical dignity , involved as the analyst's *best colleague* , in the full sense of the term." (Bezoari, Ferro 1989).

Unconscious contents are generally communicated, processed and filtered by each analyst's theoretical models, who return them via preferred "univocal choice". Weak interpretation on the contrary, maintaining a certain degree of incompleteness, refers to the patient's expressive models and through them finds the meaning to convey.

Incompleteness guarantees that the meaning is open and thus unpredictable and not conditioned by the ideological and preconceptual limits of the particular metapsychological models applied.

The non-linear processuality of the analytical field is thus facilitated; it creates new paths of meaning which reveal themselves in an apparently paradoxical alternation, since it is free to evolve according to its own unpredictable and new characteristics, despite this, closer to the unconscious truth.

Since meaning results from a hermeneutic collaboration, it follows that the concepts of container and contained are to a certain extent reversible, reciprocal and in any case, unmistakable. The discovered meaning does not acquire efficiency in itself but above all as part of a transformative process continuously opening onto new configurations. It is, in practical terms, a "semantic openness which is, notoriously a peculiarity of artistic expression" ... (...)

Freud, at the beginning of his career had already compared clinical cases to artistic expression, reluctantly admitting that they were read more as short stories than ordered scientific reports, which was due to the conversational tone of language he used during treatment (Freud 1895).

Nonetheless I do not want to argue that the narrative, linguistic, aesthetic and evocative dimensions exhaust the nature of the psychoanalytic process. This would misrepresent the profundity of Freudian thought, which describes, instead, the means of poetical art as a viaticum to the knowledge of the most concealed aspects of the unconscious.

Freud, however, had already dealt with language as an artistic component of analysis since 1907 in his work *Creative Writers and Day-Dreaming,* the subject of a lecture he had given at the publisher Heller; its uncut text was then published on the "Neue Revue", showing the literary interest the question had aroused.

In this short essay Freud assesses that day-dreams are a revival of unmentionable childhood desires.

Day-dreams had already been studied by various scientists such as Pierre Janet in *Névroses et idées fixes* (1898), Haverlock Ellis in *Studies in the Psychology of Sex* (1899) and by Freud himself, as known, in *Interpretation of Dreams* (1899).

These fantasies can be fulfilled through poetic work. This had already been sustained by Freud in a letter addressed to Fliess on January 7th, 1898. Commenting on a short story by Conrad Ferdinand Meyer he wrote: "It magnificently shows what would occur in the fantastic creation in the following years, with fantasy shifting from present to past, so that the new and old characters are on a continuum, with the former providing models for the latter." However, Freud wonders what can transform a repulsive childhood fantasy into a "poetic work "and suggests that "poets soften their selfish fantasizing by altering and veiling it, and seduce us through a purely formal , aesthetic pleasure which they offer us while presenting their fantasies. He calls this pleasure, which is offered to us so that we can release a deeper pleasure, from deeper psychic sources: *alluring premium* or *preliminary pleasure"* and he states "to be convinced that every aesthetic pleasure procured from the poets is a preliminary pleasure, and that the true appreciation of the poetic work derives from the discharge of our psychic tensions. This may also be facilitated by the fact that poets enable us to enjoy our fantasies without reproach or shame" (Freud 1907).

However, he had already dealt with this subject in *Wit and its Relation to the Unconscious* (1905), particularly as far as humor is concerned.

He claims that "humor is a way of enjoying pleasure despite the painful affects which should disturb it; it displaces the evolution of these affects, it substitutes them. (…)" "The pleasure of humor, therefore, undoubtedly originates, at the expense of this missing affective discharge, it rises from the *expenditure in affect."* (Freud 1905b)

In his work *Hysterical fantasies and their Relation to Bisexuality* (1908) daydreams are compared to night dreams , even if in a deformed and complicated shape, daydreams are therefore manifestations of unconscious fantasies, which can also manifest themselves as hysterical symptoms or attacks: Freud specifies that "unconscious fantasies have either been unconscious all along and have been formed in the unconscious or- as is more often the case-they were once conscious fantasies, daydreams and have been purposely forgotten, and have become unconscious though repression" (Freud 1908).

All hysterical symptoms, like daydreams and night dreams are partial fulfillments of a wish accompanied by an amount of pleasure which justifies their automatic repetition.

Freud has often dealt with the subject of reverie over the years; in *Some General Remarks on Hysterical Attacks* (Freud 1908) he underlines again that "if an hysterical who suffers from attacks, undergoes psychoanalytical therapy, it soon becomes evident that these attacks are nothing but fantasies translated into a motor dimension, projected into motility, and performed like a pantomime. These fantasies, apart from being unconscious, are completely similar to those imagined in daydreams or obtained through the interpretation of night dreams."

The link between play, fantasy and poetic work gives Freud the opportunity of describing the aesthetic functions of language as facilitators of the analytical process.

This subject has been resumed and developed in Formulations on the *Two Principles of Mental Functioning* (1911) as a transformation of the pleasure principle into the reality principle.

Fantasizing starts with child play, wavers in day-dreams conveyed by the pleasure principle, which survives the principle of reality.

Unconscious fantasies remain however omnipotent, negotiating with consciousness when those fantasies are artistically expressed.

As Freud states "art arrives, in its own way, at a conciliation of two principles. An artist is originally a man who turns away from reality because he cannot come to terms with the renunciation of instinctual satisfaction which it at first demands, and who allows his erotic and ambitious wishes full play in the life of fantasy. He finds the way back to reality, however, from this world of fantasy by making use of his special gifts to mold his fantasies into truths of a new kind, which are valued by men as precious reflections of reality" (Freud 1911).

This passage suggests that the analysts should use a poetic language, those "truths", pleasant, real and therefore utterable and faithful to the principle of reality.

Furthermore, Freud here applies without stating so, a complementary relationship, which has characterized many aspects of his theory, a logical antagonism between pleasure and reality principles.

He thus discovers the principle of *order from noise* later formulated by the cybernetician Heinz von Foerster (1959), a principle which is the opposite of *order from order*, that is , the order established by natural laws and leads a path to complex epistemology. The principle of *order from noise* actually expresses the concept that organized structures can originate from agitation and turbulence, which use disorder as a complementary function, not only for the benefit of their stability but also for a further increase of creative turbulence.

Similarly to every complex process, analysis as a whole tends to an increased organization, which does not mean an increase of the boundaries but also a development of new possibilities connected with the gradually emerging state.

Let us consider the case of overcoming a neurosis with multiple symptoms, via a more o less thorough analytic process.

The reorganization of the deep structure certainly leads not only to an increase in the possibilities compared to the previous structure, thanks to the range of new non-neurotic behaviors, but also to an increase in bounds since the new psychic organization can diminish the possibility of a structural regression and of a recurrence of symptoms.

The pleasure principle, reality principle, unconscious and conscious, cognitive and emotive, conflict and solution, health and pathology, reciprocally establish a dialogic relationship, according to which "two natures, two principles form a whole, without thus dissolving duality in unity, we can then speak of uni-duality" (Morin 1985).

Transforming, making the unconscious emerge, structuring an apparatus for thinking through metaphor, comprise strategies as "only a strategy can make us advance within the bounds of uncertainty" therefore strategy is creativity in action, it is art.

"Strategy is the art of using information obtained through action, of integrating it, of developing immediate action plans, and of obtaining the maximum certainty in handling uncertainty" (op. cit.).

Analytical science is ultimately an art, being a knowledge strategy. Even though, according to Freud art does not have these intentions, nevertheless it implies them as I have tried to demonstrate.

"Seen in this light the use of metaphors is valued, above all if they are not hyper codified and therefore sterile."

With Antonino Ferro we arrive at the concept of narrative interpretations as transformations which, maintaining the metaphoric nature of the patient's communication, even while developing interpretations, assure a high degree of non-saturation.

Therefore the analyst's thought does not impose a meaning, does not *colonize* the patient's text; the meaning can be found by them together, conjunctively detected in the folds of the relationship and of the transferal distortions.

Let us analyze how metaphors are used in psychoanalysis.

A metaphor can be conceptualized as a "primitive abstraction which carries a reference referring to real relationships in one situation, in order to facilitate the recognition of similar relationships in another situation" (Beck 1911).

Much has been written about metaphor since ancient times. Aristotle already speaks diffusively about it in *Rhetoric* and in *Poetics*. Since then, despite the large number of published works there have been no substantial innovations (Eco 1984).

A metaphor contains the possibility of "unlimited semiosis", of words in their comparative and communicative function.

A metaphor self-referentially relates to its cognitive description because "every specifically conceptual, cognitive operation referring to a metaphor turns out to be a metaphor of the cognitive process itself" (Morlotti 1999).

This is so because language is fundamentally metaphorical, therefore one cannot speak only metaphorically about metaphors, even if it is also true that some metaphors are condensed in prepositional descriptions such as the logical-mathematical descriptions of the scientific theories.

In light of all these considerations, the value of the metaphor cannot be denied only because of its insuperable impossibility to be fully hetero-defined.

The concept of metaphor is similar to that of the unconscious: the attempt to exhaustively define metaphors dilutes their meaning according to an indeterminate analogy used by Heisemberg.

Metaphor and concept confront each other tracing an irreversible arrow of time. In fact, "a metaphorical image, founded on an effect of retentivity, trace, that is, memory, evolves along a path over time of more or less unlikely, fantasized analogical connections, initially improbable associations and, dissociations, metamorphoses which renew the sense of the concept it connects to.

The relationship between knowledge of metaphor and metaphor of knowledge is therefore "unary", a symbol of the maintenance of an original unity of sense (...) as referring to the articulation between the cognitive rational-conceptual moment which determines the theoretical and practical functionality of the thing and the imaginative-eidetic moment giving the sense which the connection, reciprocally articulated between the known thing and the knowing subject, has for the latter. But this unity is maintained through a continuous evolution of discursive images and a regeneration of their evocative power, through a continuous collocation of new metaphors outside the conceptual system comprising the systemization and codification of the previous metaphors" (op. cit.).

Metaphor does not always have an organizational function, both in the archaic and more evolved forms of thought, for example in order to describe or imagine the basic principles of the more complex scientific theories.

Plato first contrasted the figurative, mythological language to non-figurative, factual reasoning though admitting that the first form of thought is more suitable to describe the highest truths.

Fernandez has studied some basic processes constituting metaphorical thought and which can, in my opinion, be interpreted according to both the coding and decoding of the metaphorical contents.

A metaphorical process implies that the conveyed abstraction is connected with a concrete image, referring to feelings, emotions and perceptive experience. A metaphor, moreover, implies the capability to fill logical gaps, to establish a connection between a part with the whole and even more importantly, the capability to describe non-verbal phenomena and behaviors.

Ultimately, metaphor is a process transferring information from a sensory domain to a semantic domain.

Its fundamental logic should be ascribed to perceptive laws, in particular to that perceptive quality called *synesthesia*, perception through a different sense from that used by the perceptive apparatus at the moment of the sensory input (proximal stimulus).

In linguistic terms, it consists of the description of a stimulation received by a sensory organ through a predicate referring to a different sensory organ. For example in the sentence "It is a cold color" the visual sensory experience of the color is described through the cenesthetic predicate "cold". In "what I feel clashes with what you say" the perceptive cenesthetic experience referred to by the predicate "feel" is described in a way and via an auditory predicate, "clashes".

Thanks to synesthesia, sensory domains are interfaced and connected at a primitive pre-semantic level; hence predicates like "cold" and "clashes" can be referred to polysemous referents based on the synesthetic connectivity itself.

Levi-Strauss already argued the equivalence and isomorphism between sensory codes; sensory logic has therefore been recognized for a long time, an "ana-logic" preceding semantic and syntactic categorization.

In the sensory field, forms do not comprise fixed categories and can assert themselves beyond linguistic rules, affecting the same categorial code through their own "ana-logic"; this is confirmed, for example, in combinations of colors, feelings and sounds of poetic or simply evocative language, providing an infinite combinatorial mix of expressive possibilities.

Metaphor is above all a process of synesthetic thought passing from the sensory to the semantic level of processing; the latter then retroacts on the former creating new synesthetic and meaningful combinations.

The naturally autopoietic process is a creative game thanks to the partial superimposability and (interchangeability) between sensory "ana-logic" and categorial logic, identifying from time to time the only meaning intentionally formulated.

The use of metaphor as expander of meaning and amplifier of thinkable realities is therefore quite important. In short, metaphors, being *codifying processes* themselves, maintain a transformative power between different codifications which belong to levels, the analogic and the digital-categorial.

Metaphor helps children build cognitive and affective maps of their own world, it is a tool for knowing, for effecting classifications; it is permeated by affectivity. Through it, children, from early childhood, ironize, have fun, play.

A survey by Billow, based on an experimental approach, reports that metaphors are widely used (50% of the observations) by three-year-old children, decreasing to 20% among six-year-old children; these results are confirmed by other experimental studies (Marti 1979), (Syder 1979), (Billow 1981).

These data prove the value of metaphor for learning the first figurative meanings, that is, the importance of image as an organizer of ideas and contents.

The decrease recorded among school-aged children does not indicate their diminishing capability to use and invent metaphors but rather their acquired ability to employ a denotative, literal, conventional language.

The use of metaphorical interpretation seems to be the only way of knowing the most archaic experiences of the preverbal age, they cannot be verbally represented and have no place in the unconscious but chaotically wander, fragmented and scattered, searching for the first connective solution.

Shared and co-created metaphor, has not only the characteristic of uniqueness and of the yet unrepresented, but also that of multiplicity, as the innumerable fragments of sensory-motor experiences find an immediate and unitary organization through the synesthesia of visual, acoustic and cenesthetic images often in the form of the analyst's *reveries*. At first metaphor seems to be image, percept, aggregate of trans-modal *figures*, comprising every sensory, reciprocally connected modality with the function of representing the primordial forms of thought.

The percept, produced by perception, is placed at the limits of comprehension, as confirmed by the etymology of the Latin verb *percipere* composed of *per (thoroughly)* and *capere (seize)* according to which perceiving does not only refer to feeling but also to comprehending the feeling, in a dimension where cognition, affect and sensoriality are dynamically connected.

According to its etymology derived from the Greek metaphorá (transfer) and *metapherō* (to carry over, to transfer), the word metaphor has two acceptations: one refers to "reinterpretation re-translation"; the other refers to "change" "mutation".

As confirmed by clinics, metaphor as image is able to change, to modify sensory-motor traces of archaic experiences and to carry them beyond the sensory field to the limits of the utterable, of the word.

The word is only the last step, it is the transit from image to *logos,* to the word, to verbal representation (metaphor), permitting the primitive contents to be expressed through both the rational and affective language, through both the digital and analogical representation.

> "The poetic discourse brings to language aspects, qualities and values of reality for which there is no means of direct descriptive expression and that can be uttered only thanks to a complex language game between metaphorical utterance and transgression of the common meanings of our words. Metaphorical utterance would have the capability to re-describe a reality which cannot be directly described. (...) "To see how", which is the metaphor's strength, would reveal existence at the most radical ontological level." (Ricoeur 1983)

Through metaphor the cognitive system is able to grasp what is "similar to the thing" (primary sensory experience): the analogical, with that which economizes its discursive manipulation, and the digital.

Moreover clinical experience shows the deep connection between metaphor and reverie.

Reveries do not necessarily manifest as images; they express themselves as humoral states, body feelings, slumber states transiting quickly to consciousness, aggregating residuals of perceptions and primitive globs of meaning suggested by the listening unconscious. They are personal intersubjective events, originating from the meeting of emotions of transfer and counter-transference, they are jointly created: they are the "voice" of the intersubjective analytic third that "originates in the dialectical relationship between the subjectivities. The analyst and the analysand together contribute and participate in an unconscious intersubjective construction (the analytic third) (...) specifically, the relationship of the roles of analyst and analysand structures the analytic interaction in a way that strongly privileges the exploration of the unconscious internal object world of the analysand".

> "The experience of patient and analyst in relation to the intersubjective analytic third is asymmetric not only in terms of the way in which each contributes to its construction and elaboration. It is also asymmetrical in that analyst and analysand each experience the analytic third in the context of their own, separate, individual personality system, shaped and structured by their own form of psychological organization, their own layers and linkages of personal meanings derived from the totality of their history and unique set of life experiences, their own modes of organizing and experiencing bodily sensations, and so on. In sum, the analytic third is not a single event experienced identically by two people; rather, it is a jointly but asymmetrically constructed set of conscious and unconscious intersubjective experiences in which analyst and analysand participate" (Ogden 1997).

Reveries can also be thought of as the product of unconscious receptivity, when it is free to receive the experience with the minimal mediation of the secondary (conscious) process. The psychoanalytic setting which uses a coach is a facilitating context for their appearance, but it can also arise following other ways, in other places and times, in an apparently chaotic sequence of allusive and evanescent traces. These productions have affinities and differences in comparison with night oneiric phenomena. The latter are mainly images poetically defined by the laws of the unconscious such as condensation and shift and circumscribed in a known neurophysiological activity. The former occur in every state of consciousness and the range of their expressions is incredibly wide.

A stomach ache, a misunderstanding about a session start time, a muscular cramp, feeling enticed by the recollection of an apparently insignificant experience, having a fixed stare on a detail in the room furniture, an automatically repeated gesture, a scene of a recently seen film, the lines of a poem which one has not enjoyed for a long time and infinite other experiences can be the result of an incredibly articulated unconscious activity, and can enigmatically arise through the introduction of meanings which cannot be verbalized yet. Rather then organized dreams, they seem *contrariwise nightmares,* fugitive explosions of thinkable elements, primitive states of aggregation accompanied by the activation of the autonomous nervous system, "fantasies in the body" (Gaddini 1982) begging not to be removed.

In general, reveries require a long work of revision, discussion, supervision, before being formulated as verbal, though metaphorical, interpretations.

After being transformed and interpretatively codified, they can be the first meaning of experiences never before thought ; they have not even appeared in the unconscious, scattered but still present in that shapeless, protomental area connected with archaic, psychic, psychotic events or with events referring to abandonment experiences so early that they cannot be metabolized. The transit to utterability can be very hard. Distant narration, metaphor is sometimes the only way of signifying reveries in the communicative act.

Reverie is actually a precursor of metaphor, suitable to generate meaning. Its primordial, pre-verbal form is already permeated with meaning like a pictogram (Aulagnier, 1975); even if it is not yet utterable, it is at once dream and hallucination, delirium and striving towards words.

It can become an "analytic object" since it is suitable to be used to create analytic meanings; but its archaic nature places it on the border between soma and psyche" (Bion 1962), (Green 1975).

It conveys the directness of rough contents which seem to differentiate it from the more evolved oneiric products; yet it refers to dreams, sometimes using its images and contents as *night residuals* the same way as dreaming uses the mnestic underlying traces of the daily activities.

Thanks to their protomental characteristics reveries can also be compared to "free associations", a felicitous expression of Ferenczi. These are gestures, postures, involuntary muscular contractions, dizziness, which are part of the patient's behavioral repertory during analytic treatment.

Psychoanalytical body therapies underline the multilevel position of these phenomena which talk about and for the unconscious. There are motor, sensory, emotional levels, which are often imbricated, leading us to think that the α function (the Ego function which organizes and gives meaning to sensory input, according to Bion) does not only operate in cognitive thought but that there must be an α function for sensation, emotion and body motility (Downing 1995).

Therefore reveries are composite, condensed processes, concrete manifestations of intersubjective relational events whose comprehension makes possible the integration of proto-thoughts which make β elements thinkable. They are precious tools at the service of metaphorization.

Now let us see two short examples showing different clinical applications of reveries as metaphors which promote non-linearity and organization of meaning.

Hansel, Gretel, Giotto's O and Swing

The analytic relationship with Paola presented problems since the first session. Her defiant and suspicious attitude, her evident verbal aggressiveness are accompanied by an uncommon intelligence, a complex symptomatology consisting in serious mood alterations, instability in affective relationships, quite frequent induced vomiting and a penchant for deceitfulness in friendship.

Despite the difficulty of containing her hostility, to bear her continuous provocations, and her frequent skipping sessions without informing me, in the meantime it allows me time for thinking. After a long consolidation of the analytic relationship, interpretations concern the conflict between "vomiting" out the bad object and at the same time desiring the attention of the conquerable mother, as Paola sees it, processing her mother's mourning experiences caused by her sister's death in early childhood.

In the transfer I am assimilated now to the bad (rejecting) object then to a good object, uncontrollable and consequently unserviceable, rather threatening; due to the patient's express wish for independence.

Her claims for independence are so insistent that I expect her all of a sudden to give up therapy.

Projective identifications create an atmosphere of fusion where I act as a partially distinct object, good enough to permit a primitive introjection.

That partially satisfies Paola's greed but it, at least, allows me, under strict control, to be admitted to her internal world which till now has been inaccessible.

The relationship gets better and the patient shows clear signs which attest to my settlement as good object.

Paola fantasizes about my reassuring presence on the chair next to her bed; she explicitly assimilates me to her loved grandfather and her aunt with whom she has a good relationship.

I gradually become the idealized, totally good object. Some days after a significant session, while I am speaking with another patient, the discussion touches on the fairy- tale of Hansel and Gretel. An image immediately arises from my thoughts. Hansel is passing the bad witch a bone to convince her he is not fat enough to be eaten yet. I associate the reverie with Paola, to her refusing food and the image fades.

It comes back again and again on the following days, while I am waiting for the patient's associations to help me (us) understand.

The image of Hansel is quite pregnant with meaning charged. I feel it vividly, and the meaning gradually emerges.

I share the image with Paola. She immediately invites me to play a closed-eyes game in which we have to say what else we observe in the selected image. I accept the manipulative game without interpretation.

I start. It is dark; I see a cage, with a leaning hand holding a bone. Paola asks me if I can see Hansel. I answer that I do not think so. Paola continues: "I can see him; he is chubby, surrounded by food he hides. I can see Gretel as well, she too is hiding It is dark, there is nobody..." and goes on describing the cage as she sees it: It is this element which now is standing out in Paola's associations as in my still shapeless, emotive echoes: the somehow evocative fulcrum produced by the analytic third.

My mere mentioning "the cage" in a questioning tone, now produces the effect of an incomplete interpretation, of a metaphorical access to a rich series of reveries and associations which are leading Paola to the limits of utterability and which I am going to put in words.

Paola says she feels like Hansel and Gretel at the same time, she is the one who controls the bad witch, providing her with only the least desirable contents. She has uncomfortable feelings about the claustrophobic cage; she makes associations about the ideas of freedom and independence.

I interpret her difficulty of wriggling out of a primary object experienced as "encaging" and I underline that she still may still need to control me so that I do not "encage" her. "I do not want to get sucked down" she states. "By your own greed placed inside me" I answer. She opens her eyes surprised. A paired sequence of shared reveries has thus developed and defined a field which played at encaging (sucking down) both of us in an unthinkable spiral.

The opening move, Paola's suggestion to play with closed eyes to control the analysis, has now acquired other values; it has become a common method of investigation , thanks to the fact that I avoided a premature interpretation which would attack the heart of her defense but then bounce back against me with little hope of development; it had also created an area where non-linearity (unpredictability) had proven to be an explicative and creative element highly refined by the "encaging" and "sucking down" metaphors.

They alternate to find the meaning appearing before them with the naturalness of a "real" game, where confusion replaces sharing in fueling the intermittence of a coupled entity continuously tempted to slide downwards toward the unsaid, toward the un-thought, toward the omnipotence of a cage-crib-casket which awaits both.

Reverie with Antonio is biunivocal and simultaneous, mostly expressed through images. After one of my interpretations, the patient tells me that he is thinking about something else and that he cannot follow me. I see a decided gesture of a hand tracing Giotto's O in front of me, slightly on the right. Antonio at the same time tells me he can see himself in the background, on my right, while he is doing a golf swing which he defines perfect; a unique, mechanical gesture, in which the ball is one with the movement, a chance accident along the uncorrupted trajectory of the gesture.

Two images, two people, a single visualized subject. The swing as perfect as Giotto's O, I comment is the round, warm and omnipotently out-of-time place where the patient takes refuge to avoid the fear of death, fading due to my impertinent interpretation, experienced as a tearing attack from which one has to escape. It is the desperate attempt to find consolation in an autistic dimension which must ignore everything to protect a fragile and boundless Self.

Making every gesture perfect, returning compulsively to the image of the swing, avoids the thought I try prematurely to make prevail. It may be also an attempt to circumscribe an outline which can serve as a protection of the threatened Self, a primitive psychic skin, containing something still shapeless, barely sketched.

The use of reveries as metaphors-interpretations represents the triumph of uncertainty as a creative act par excellence, as a favored viaticum to narrative and intersubjective truth, in other words indissolubly connected with dialogue and research done between and through observers.

It is in essence a "personal mythopoeic construction of the analytic couple" which can be most clarifying as a shared representation close to the particular text of the single subject "which constitutes a, subjective and unique reality as unique as is the relationship between the analyst and the patient" (Arrigoni, Barberi 1998). Interpretation thus becomes rather literary since, similarly to modern narratology, the patient is asked to actively collaborate in the construction of meanings.

Therefore there can be a typically metaphorical interaction between two people speaking at one level but communicating at another which is closer to the relational sense.

This results in an increase of creativity rendering vital and original the text on which the analytic relationship develops according to highly unpredictable dynamics.

In other words, narrative interpretation promotes the non-linearity (plurality of possible solutions) of the transformational process substantiating the analytic field, leading us along new paths ever closer to relational truth.

This aspect is, in my opinion, connected with the capacity metaphor has to express a very similar concept. "Metaphor is an indicator of the non-linearity of text or of thought, it is an indicator of the openness of the text or of thought to different interpretations or reinterpretations, leading to a personal way of discussing the ideas of a reader or an interlocutor"; it indeed "overcomes the discontinuity and isolation of things", expressing "nuances which merely objective and denotative language cannot produce", similarly to poets who, as René Descartes reminds us, "exploit the power of image" (Knyazeva, Kurdymov 1994) (Morin 2000) (René Descartes).

Chapter 35

PROMOTING THE NONLINEARITY OF TRANSFORMATIONS

The transformed original, invariant of the transformation process, belongs, according to modern psychoanalysis based on the new epistemology, to the truth of the field rather than simply to the mind of the patient, as if they were a mere object given to be studied by a neutral observer.

The truth thus is revealed in coupled truth, a fact which increasingly regards the relation, given that the object is not completely objectifiable, separable from the observer.

That is where we have to start in order to study, with the contribution of complexity, further modalities of transformation, further methods to work in complex processes to promote their non-linearity, since Bion's transformations into O are complex transformations.

This is still only a theory, even if a joint study on the relationship between psychoanalysis and complex thought could be operationally promising as well.

Complex epistemology is indeed an epistemology of standard procedure, it originates from the concrete observation of the phenomena taking place; it is not aimed at a new metapsychology, but rather at a new approach to something already known.

In the perspective of *functional mobility,* even the applicative contribution of complex thought to clinics will be able to increase; other modalities of observation of known phenomena will likely be found, as well as new ways of promoting the change, that is, the nonlinearity of transformations.

Chapter 36

INTERACTIONS

There are two ways of considering things, as objects and as signs.

St. Bonaventure

The complex approach to psychoanalytic concepts is a field of abstraction where the concepts are in a certain way scotomized, in favor of a description focused on the relationship, on formalization and on the concept of transformation.

Bion already laid its bases even if not from an explicitly intersubjective point of view.

According to Bion algebraic formalism is a pure method of transmitting knowledge, a justifiable and clarifying aspiration. However this does not mean that it would be desirable to derive psycho-affective processes in a totally abstract way, cut off from the contents; on the contrary it seems convenient to retrieve "what is related" avoiding the reifications inherent in objective description.

Words, like other, non verbal, means of communication have the power to evoke content essential for the comprehension of psycho-affective processes; so long as they are not an imposition, a fixation definitively preventing role play by causing confusion and misunderstanding via the illegitimate, arbitrary overlapping of meanings.

This happens more often than we can imagine, mostly when we consider the other person's communication impersonally, decontextualized from their suffering and their experience.

In other words, it is necessary to stay in contact as much as possible with the ambivalence of communication to assure that information about the other person (of the other person) take form, even at the risk of losing meanings already attributed.

If we deeply analyze our way of thinking we will find a persistent tendency to homogenize similar meanings in the same way as we assimilate reality to our meanings.

On the contrary we have to maintain a dialectic so that contents and surface can be disambiguated.

These considerations lead us to consider contradiction as the real driving force of thought, able to make it develop strong and healthy, far from a presumed a priori knowledge.

As Morin reminds us "in any research *it is necessary* to understand the middle through the extreme, equilibrium through tension, the measured through the uncontrolled, the normal through the pathological" (my emphasis) (Morin 1969).

This is the more appropriate way of working through our own projections, being however aware of their nature and of the fact that they need to be shaped through the disaggregating tension of doubt, of emptiness, of the other person's truth.

Our speculative behavior will thus also change. We will acknowledge the uncertainty of what is communicated in the analytic relationship and we will be freer to address ourselves to the evaluation process through further clinical material and through the counter-transferential answer.

Chapter 37

NAMING

Morin underlines what is still valid in psychoanalysis at the moment of the interpretation. "It is necessary, as a general rule, to let them express themselves, show themselves, but at a certain point, we must give a name" (Morin 1969).

This interpretation takes place thanks to a split of the analyst, who on one hand empathizes with the patient and on the other hand acts as an autonomous organizer of a text; taking into account the information they have been given.

This is the creative and transformative function of the analytic work which widens our understanding in sight of new horizons of meaning.

This transformation leads to sharing thoughts via the open expression of innermost feelings, to the cost of an omnipotent cognitive pretension at the service of analyst-patient fusion.

The patient's account is a description of their affective contents; the analyst's account a re-narration. In the final analysis, a new reading (re-naming) of the text given by the patient, introducing the symbolic regime, making the communicated contents meaningful and, similarly to narratology, corresponding to a semantic isotopy .(Arrigoni Barbieri 1998).

From a perspective of complexity the therapist can get closer to the truth if they accomplish a further decentralization or an observer split (on the part of the analyst), in order to analyze the coexistence, in the contents communicated by the patient, of convergent indexes (consistency) but favoring divergencies according to a principle of competition in the evolutive nature of the psychoanalytic process. Like every complex process it is a result of competition between features of order and disorder in dialectic balance.

This is certainly a radical strategic choice, imposing doubt as a legitimate driving force of thought.

As Morin reminds us, in a naming process based on complexity, criticism must go to extreme consequences "as things stand now, it is a matter of transforming the principle of coexistence of the different sources of information - information in the strict sense of the word and the interpretations of the interpretative systems themselves" (Morin 1969).

Therefore it is necessary to constructively criticize even the interpretative systems themselves, the various metapsychological models.

Psychoanalytical knowledge and complexity are thus in a relationship where concepts recursively and dialogically shift from one field to the other, reciprocally nourishing and providing a meta-point of view which legitimates their connection.

A new horizon opens up to new knowledge continuously self-nourishing through a synthetic-pluralistic methodology, autopoiesis of the cognitive process itself.

"Who will find the Sinaic site from where we can embrace together Elohim and JHVH; Chaos the destructor-constructor and the Principle of the Law? Who will write the new commandments?" Morin asks himself in a heuristic attempt to get to the heart of this matter (Morin 1969).

Chapter 38

EPISTEMOLOGY AND STYLE

No wonder St. Matthew looks so apprehensive in Caravaggio's painting of him, clutching his pen while a rather thuggish angel dictates to him what he must write down: the act of writing come weighted with a burden of anxieties. The written word is so much like evidence-like something that can be used against you later.

Margaret Atwood

'If you give me a copper coin I will tell you a golden tale.

Robertson Davies

In this appendix I will describe briefly how, in my opinion, the new epistemology of complexity affects the style of clinical reports by one psychoanalyst: Antonino Ferro.

Ferro's particular style is obviously not only the result of his epistemological choice; however I underline that it is important and concerns, through the position of the subject (analyst-narrator), propositions, the rhythm and color of the narration, and reference to an observer (the analyst) the inseparable participant in the narration.

It is a narration with a pressing rhythm, in order to express the nature of the interpretation which "gropes in a net of evanescent references, between an oral test, evasive and dynamic, non objective codes and unknown referents" and the unpredictability of an autopoietic and non-linear evolution typical of a complex description (Arrigoni, Barbieri 1998).

Despite that it is possible and necessary to organize the narrative so that it can be used by the scientific community.

"If we agree with Benjamin (1995) (...) that things speak without being organized, then we can think that the idea of writing a clinical report arises in the analyst's mind when the temporal dimension of the analytic work based on the past and present is enriched with the future; when, we could say the archaeological metaphor joins the metaphor of the journey, offering resumptive and integrative possibilities which indicate the structuring , in both the protagonists, of a four-handed sonata, of the final movements" (Sarno 1990).

I have chosen some clinical reports written by Ferro which are, in my opinion, illustrative of his style and of the implied and implicit epistemology, trying to define its mark on the writing.

Also the analyst's cautious attitude in formulating their own interpretations is reflected in the author's choices and refers to a multi-vertex perspective in psychoanalysis.

Note, for example, how Ferro and Bezoari move from theory to clinical description, with no cuts, in the most natural and clearly descriptive way, indicative of a mature turning point in their theoretical approach.

As if they were talking with the reader, they explicitly say: "When during a session , we exceed in applying our theories, our patients are, as usual, prompt in pointing it out , for example through dreams , in which they are given old records or they meet civil servants reluctant to renew their expired identity cards" (Bezoari, Ferro 1991).

In this passage theory is simply introduced directly old records and other oneiric images are put on scene as functional aggregates emerging from the minds in relationship, relationship which is precisely described, with so many allusions that no further theoretical explanations are necessary.

Moreover consider how the clinical case of Nico can easily be inserted in the discussion about the risks and the benefits which the transit of projective identifications produce in the analyst's mind and how the clinical narration itself quits hesitating in the conclusive argumentations.

> To conclude, we would like to propose some food for thought about what fieldwork implies for the analyst's mental health, according to the models defined here. .
>
> If we accept that permitting to be moved by the patients' projective identifications, although within certain limits, is an essential component of our therapeutic function, this involves being exposed to mental suffering whose intensity and quality are not predictable from the start.
>
> Here is a crucial question: what risk are we willing to take in order to accept and metabolize the unknown experiences which the relationship with the patient can activate in us?
>
> Nico is a seriously disturbed patient, he is living, maybe for the first time (as the end of the therapy is approaching after seven years), the painful and for him excruciating experience of separation, which he feels and fantasizes as an agony, a death sentence. During a session in which he is painfully expressing these feelings, (which he has never experienced before even though he has undergone traumatic detachments from his mother) the patient says that he is smoking a lot in this period, more than 40 cigarettes a day, even if he knows that it can ruin his health. Suddenly he asks if he can smoke at least one cigarette with him. Rather puzzled the analyst answers with an interpretation, thinking that if it is correct, it will solve the problem, helping him out of the difficulty; he says something about his wish to overcome the anger and hate (previously emerged) for the analyst by whom he feels he is going to be abandoned, through smoking the peace pipe (the confidential language of the couple already included "far west" metaphors).
>
> Nico nods but he insists and passes a cigarette he has already lit. The analyst cannot avoid taking it. Then taking advantage of what seems a moment of distraction for the patient, he drops some ashes in the ash-tray. Immediately Nico comments: "Doctors want to keep healthy, they do not smoke....I only have to smoke a lot" Incapable of understanding; worried to cause a short-circuit reaction, the analyst feels he has however to take the risk and he is going to put the cigarette in his mouth when he has a sudden panic attack: the possibility of infection, the specter of HIV.
>
> Only then, after overcoming this moment of terror, the analyst feels to have actually been in "O" with the patient and with what he was asking him. That is to witness his suffering , keeping safe as if he were behind a window or to have enough willingness and courage to share in real life his terrible and painful experience, taking the risk, even though on a reduced scale (a pull on a cigarette against the patient's 40 cigarettes). The analyst verbally expresses

these thoughts. The patient answers: "I have been wondering about that for years". Even if the guarantee of the safety net provided by the setting is still fundamental, however it cannot be considered either total or always efficient, as those sessions of off-hour mending, the counter-transference dreams, demonstrate (Bezoari, Ferro 1991).

This passage is not simply an example of expositive elegance, it constitutes a node of a plot without interruption between theory and clinic, a real distillation which privileges a metacognitive perspective, which, at the same time is not separate from the object of its knowledge but is immanent.

In that plot, the knowledge of the object itself becomes cognitive autopoiesis, concurrently same and different making contradiction sustainable as an integral part of the heuristic moment innate to our way of thinking. Metathinking does not therefore imply thinking from an alien or hypostatized, infinite point of view, but rather an integrated movement of thinking the knowledge in the knowledge object.

Subject and object are distinct and one, in a shared, even, emotional pursuit of truth. How is it possible not to be infected by so practical a theory and by such a theoretical procedure of clinical description? The effect says a lot. Let us report other passages by the same author taken from more recent works.

The Patient's "Dialect"

I like to make love with my husband provided that his penis penetrates me by 3 centimetres, not less because I do not feel anything and not more because it hurts....

What is this patient talking about? About her sex life with her husband and the forms it takes. Or, we may say, about reliving forms of breast-feeding in the transference, if we consider a geographical confusion and the difficulty of finding the right degree of penetration. About the mental functioning of the couple in the analyst's consulting room in which the analyst's words must penetrate with the right degree of penetration, not less because they will not reach their destination, and not more because they are persecutory. However she is also saying that a container has already come into being and that this container can withstand so much stimulation but no more.

She is saying all these things, and much more besides, if the analyst is able on that particular day to abstain from decoding according to his prevailing model and to venture to share the "dialect" used by the patient in telling her story and to learn more about it - what happens in making love; what happens in inadequate or excessive penetration - and to wait for the possible scripts and sets of these penetrations to take shape. The aim will be to allow the analyst-patient relationship to generate an increase in meaning rather than leaving the degree of meaning as it was (by translation into another dialect that is more confortable for the analyst) or positively impoverishing the meaning (mortification, stiffening or skeletonization of the communication when read by a rigid code: transformation into hallucinosis).The more seriously ill the patient, the more necessary it is to increase the degree of meaning, for this process is ultimately indicative of the success of the symbolization process.

The patients with the severest pathology are, in fact, those who cannot tolerate the impoverishment of the meaning of their communications. This is one of the stumbling blocks of criteria of analyzability; the problem has to do with the defenses I must erect in order not to expose myself to excessive suffering.

> How far am I able to go to meet a patient's emotional demands?
>
> To what extent am I available to receive, transform and re-narrate their emotions (as well as mine, as activated by them), rather than defending myself with the lead apron of codified theories - even the most sophisticated ones - of the kind that could be accommodated in column 2 of Bion's grid?
>
> Again, not all patients put us to the test with such primitive needs (criteria of analyzability often serve to select patients who confirm our theories). Even when, as so often, we misguidedly disregard the patient's dialect and compel them to learn the analyst's (which will be that of repression, part-objects, the unconscious body fantasy, the container-contained, etc.); nevertheless, a relative and marginal process of transformational re-narration of the patient's emotional states will take place (for even if the analyst is unaware of it, these states will pervade him, or else, if blocked off, they will be retold by the patient when not having been favorably received) woven into a more or less mythical "historical" reconstruction of early or not so early childhood vicissitudes and fantasies.
>
> Naturally, where the most primitive parts of the mind are concerned, these unconscious stratagems cannot suffice, failing a process of symbolization in the here and now mediated by the entire arc of communication (from acceptance of the emotional wave via transformation to subsequent telling-in-a-story) and leading to the formation of α elements- i.e. "thoughts"- the result will be those valuable indicators of a dysfunction in the analytic relationship that we call negative therapeutic reactions, impasse and psychotic transferences. This betokens a mismatch between the analytic instrument, the minds of the protagonists at the moment, and the analytic task in hand.
>
> This means not that we should expose ourselves to new and greater difficulties, but that we should be aware that we can advance only a little way beyond our own analysis and the theories that underlie it, and that we ought not to shirk this task.

In this passage some lines of clinical reports are so packed with psychoanalytical speculation that we cannot avoid thinking of the hologram as a logical meta-model from which to infer the vast reference in Bionian meaning. Metaphors as mortification, hardening and skeletonizing are so pregnant with meaning, if experienced in context, as to highlight at once the concept of transformation in hallucinosis, without saturating the meaning.

In general in Ferro's narrations, the reference is holo-directional from the part to the whole, from the whole paragraph to the clinical description and the *semiotic aggregator* is the metaphor which, far from proposing a boundless semiosis, enriches with meaning precisely because it refers to and describes the uniqueness of the experience.

Metaphor nourishes with emotion and emotionally actualizes the described relationship, offering new suggestions to the reader, who thus makes an entrance and imagines possible developments of another story which they would live if they could, for a moment, actively take part in the transformations happening.

Intrigue is sure; we would like to enter the text and continue it, enticed with the uniqueness of the moments the author makes us experience, and to imagine their continuation.

In order to distinguish ourselves from the author we almost have to recover, return from a journey which drags us into the twists and turns of a personal adventure and at the same time in a rigorous but not exhaustive description.

Thinkability is expanded in Ferro's description; he proposes an extension not only of the clinic but also of the theory, in order to detect the untold. We are left with a sensation of openness, of a wide ranging thought, of freedom which permits us to love our personal dialect. They must be contextualized to be living, and what first seems a debunking becomes an opportunity to understand the vastness of the psychic dimension we serve; so wide that we must decide our reference models again every time we move to new contents.

Naturally, it is not a matter of theoretical syncretism or of descriptive anarchy; or of an articulated system of thought leading us to a mindset willing to reverse our perspectives.

The very titles of the paragraphs dealing with the clinical cases show a use of metaphor which anticipates the content and summarizes it without ever separating the clinical presentation from the theory. In the following passage, taken from the above-mentioned work, metaphors are the essence of the description which reveals, point by point the shared couple experience. Following is the part I consider most significant.

Claudia's Washing Machine: Which Program?

(...) After a short silence the patient says: "I remember a dream: I was washing some of my sweaters with the extra-delicate program, which is fine for wool, but as the spin program was also extra-delicate, the sweaters remained saturated with water and I was afraid that they would be too heavy on the washing line."

I ask whether she thinks the normal wash would ruin these sweaters; no, she says, because they have now been washed several times with the extra-delicate program and there is no longer any risk of their being spoilt. She adds that she has been thinking of her sister Luisa, whose little blanket had once been hanging up after being washed, but was unusable until it was dry; that had made her cry a lot.

At this point I feel able to give her the normal wash cycle (which was expected and desired), and I bring into the transference feelings about the first film, the second film, the experiences of the "sister" who cannot wait and her new capacity to contain her emotions.

The patient replies: "Now I understand all the rage I felt yesterday, quarrelling with everyone ..." and I in turn understood my own rage at the postman who did not deliver the mail punctually which had been my way of receiving the patient's projective identifications.

This is a remarkable example of weak interpretation expressed through metaphors, in order to promote the highest level of non-linearity in the transformation leading to insight. The respondence is immediate and, by increasing knowledge, expands thinkability.

Theory is now so integrated with clinical description that in all the reported sequence it is no longer brought up thus creating an intimate atmosphere and cognitive harmony with the author at once descriptive and creative. The passage is not a theoretical reasoning but a truly fertile mutual enrichment between clinic and research.

Even if Ferro is absorbed in the dynamic, intended by the point of view of complexity, he does not mention it; his task is principally clinical , it does not imply "thinking about" but "thinking with", and differs from the work I am proposing, in which the former prevails.

Functional aggregates, characters, the analyst's understanding of them are used as various literary devices, simple and direct, clearly revealing how the unconscious functions.

Here is an example of the use of synecdoche and metonymy, in addition to metaphor.

Basque Terrorism

Mimmo comes to analysis because of a vague malaise that makes him neglect his studies, let himself go, get bored and often complain.

When I see him at the door in his grey suit, I think: "What a boring conformist he must be"; he seems me to be the last survivor of another age...and then I notice a kind of unexpected gleam in his eyes, which makes me think: "Or perhaps not...he looks like a savage".

At the beginning the analysis is difficult, with long silences: I myself am afflicted with boredom and sleepiness; I feel that something is being put to sleep but I am not sure what, nor can I find any opening towards something more living and vital.

So we proceed, until something very unusual occurs: one winter evening, as a thunderstorm is beginning, the lights suddenly go out. I am not equipped to cope with such an eventuality because this has never happened to me in all my years of practice. Left in the dark, I am invaded by unspeakable terror which I cannot describe to this day - absolute panic at the terrifying idea that Mimmo may jump on me, kill me, stab me, and rip me to pieces. Images of extraordinary violence take possession of my mind… meanwhile Mimmo drones on in the same monotonous voice...and then the lights comes back on ...the session continues...but inside me are these scenes that have opened up, but which I do not know how to use: I decide to wait and see, but feel very tense. A few days later, my attention is attracted by a "basco (a beret)" that Mimmo has begun to wear to his sessions, which is completely out of keeping with the rest of his attire, and when it falls on the floor once as he is leaving , I find myself "picking up his basco (a beret)".

At this point I have an intuition that enables me to connect what I had experienced as well as a tendency I had noticed for my interpretations to be insufficiently modulated with what was happening and I say to myself: "It is indeed the Basco (a Basque) that needs to be picked up". In the next session I cautiously allow this character to take the stage: perhaps the basco (basco (a beret)) that falls down, and which it is perhaps up to me to pick up, may it not be a Basco (a Basque) of which I have never heard.

From then on develops a whole story about the Basques, the importance of the iron (ferro in Italian) mines to their economy, and the particular explosiveness of their character - even if a young Basque has recently married a cousin of his of whom he is very fond; he has even conceived an interest in their language which appears not to belong to any known family...and then there are the bombs...the Basques' need for independence...the crushing of the Basque identity...and then in successive sessions, via accounts of films, the American bison...the wild animals from a recent trip to Africa, where his father has unexpectedly started an import-export business...and ultimately the drama of the Albanians and their needs.

Here is an extraordinary use of dots which create a pressing rhythm in the clinical narration, which involves the reader.

Clinical report and narrator take the scene offering themselves to be re-interpreted by the reader who is thus deeply involved, helped by literary devices too, which suggest the implicit and latent meaning of communicative exchanges.

Ferro insistently underlines the importance of the ability to grasp the autopoiesis of the emotional text; that is why he often does not decode it, but rather points out the signals it offers, so that not to cut the evolution of the stories which more urgently require to be narrated, as happens in the following passage.

ROSA AND THE PROLIFERATE DISEASE

An initial interview with Rosa, a 25-year-old philosophy teacher, is devoted to the feasibility of a proposed analysis, which may have to be postponed for several years because I have no time.

At this first meeting, Rosa succeeds in telling me something very dramatic that happened to her on a journey; although it will be a long time before she can commence her analysis with me, she is very determined and decided to wait for me, because she had unexpectedly managed to tell me about "that particular thing" about which she had never before been able to tell anyone; I in turn give up the idea of not accepting her for analysis owing to the long time involved, and accept her proposal. The following stories appear during the second meeting. After our talk, she "had very regretfully" had to leave the school where she had felt good. ...then a relationship had flared with Marco, another teacher at the school, who had lost his head, and become terribly involved; meanwhile she had also had an affair with another colleague Aurelio, which she described as a relationship "the way you want me" in which she liked to support him in everything; then there is the boy friend who guarantees her a secure and reliable position , but does not satisfy some of her requirements; finally she produces her younger sister with a "proliferative" disease, which has an urgent need for appropriate treatment. How are these "characters" to be thought of? One possibility might be to accept them according to their referential prevalence, as characters from external reality whose importance lays in the emotions and feelings they activate in the patient.

Alternatively, they could be seen as characters who combine and saturate values of Rosa's internal world, thus immediately permitting the step from the external reference to a theory (of internal objects), and hence to Rosa's internal groups and their organization. In other words, the characters evoked in specular form would have to do with Rosa's internal objects, unconscious fantasies and fantasy formations.

It soon becomes possible to recognize these fantasy formations in the transference and in the relationship that arose from the very beginning of the first interview. These "figures" could be seen from a variety of viewpoints, as transference entities which were thus projected, or as anticipated visions of the transference relationship in the analysis (in terms of transference based on repetition, transference based on externalization, relationship in the form of something unique and specific of the two minds in the here and now, which is a function not only of the transference, but also of the receptive and transformational capacity of the analyst's mind).

These aspects could or could not be interpreted in accordance with a whole complex of technical considerations.

From another viewpoint, however, one in a state of necessary oscillation with the others, these characters could be seen as the syncretic and pictographically narratable expression of the emotional facts occurring in the consulting room, in the mutual fantasy formations activated in the bipersonal field (...).

Instead of interpretive decoding (which would then have generated other possible texts) I opt for an open contribution to the narrative development and say that it seems to me that there are different stories here: one of them passionate, one erotic and another concerned more with affects (...). If due attention is paid to the signals of the emotional/narrated text, stories that need to be told and transformed will not remain suppressed, while all the possible stories not relevant to the current emotion and urgency may stay concealed.

Consider another example of metaphorical restitution, this time introduced by a simile which becomes autonomous metaphor.

Mario's Lifebuoys

Having had the traumatic experience of losing both parents in a car accident when she was a little girl, and not yet capable of affective cathexis because of the resulting terror, Ludovica is able to tell all this, and other things too, through a character, "Mario" (with whom she had some kind of relationship), whose characteristic feature is that he has lots and lots of affective stories, none of which can become either prevalent nor significant. One day she says that the people with whom Mario has these stories are for him like lifebuoys to cling to.

I respond "So he is like a survivor of a shipwreck who is so afraid of risking another sinking that he prefers to keep his head above water, and paradoxically in safety, by a system of lots and lots of lifebuoys, rather than sailing off in a new ship".

One the one hand , this interpretation defines a possible story (a possible narration or transformation) so it is saturated, but on the other, it leaves the patient free to accept it, reject it or accept part of it perhaps in a way we might not expect.

Furthermore, it is the fruit of the relationship; arising like a reverie in the session, it is therefore the offspring of waking dream thought, drawing on a model created without memory or desire, and not resulting from the use of saturated theories in the session. It may be a prelude to transformations into "O".

Notice how in the following passage an entire clinical story seems to be condensed in an image as pregnant as it was rapid.

The Cup from Beauty and the Beast

I once had a little girl patient who was so terrified of aggression that, playing Beauty and the Beast in the session not only on no account could she take the part of the Beast but also found it impossible to play the role of Beauty; she could only participate in the story as the "cup" in the Beast's house. That is what we must be able to tolerate: that our patients are able to express their tolerance to aggressiveness.

In *Supervision in Psychoanalysis: The Sao Paolo Seminars*, Ferro shows all the transformative potentiality of his remarks proposing reveries in the shape of metaphors which promote non-linearity in the supervised interlocutors. His references to films, literary works, memories, anecdotes and images are examples of transformative texts which cope with the unpredictability of development making it open to new meanings even if they are already present in the deep emotions of experiences.

Let us observe how in this passage Ferro promotes with great simplicity the possible worlds of valuable meaning from the text, supplying an example of non-linear, transformative facilitation.

Thematic Seminar I

Perhaps you remember the beautiful drawing by Escher, entitled "Relativity", where there are various stairways and one doesn't know where they are going and how they are connected. Characters can be located on different stairs, in different ways. On one stairway we have the character as a real occurrence; on another the character as the patient's internal

> object, and on another, in my opinion the most interesting one, the character as something specifically concerning what happens in the sessions (...). I have the impression that the patient's work within the session, or inside the analysis, consists in stopping time: He performs his job by not coming to the session. This is his work: He has to stop time. Perhaps you remember the Bergman film "Wild Strawberries". It opens with the dramatic scene of a funeral and a clock with no hands. I think that this patient's work consists of constantly removing the hands from the clock; he is a busy man - engaged primarily in avoiding thinking, like an hyperactive child who has to keep running this way and that, so as not to fall into a deep abyss of despair. At some point, the patient says this, when he talks about paralyzing time, life, his age: This reminds me of the tale by Edgar Allan Poe, "The Purloined Letter" where the best way of hiding something is to make it very obvious. (...)
>
> I thought of the depressive movement we find at the beginning of Dante's Divine Comedy: "In the middle of the journey of our life I found myself in a dark wood" -which is the way of representing the impact of middle age. The advantage is that, in the past this happened to people at the age of thirty, now it comes much later. Elliot Jacques makes the point that the only way one can work through and overcome a midlife crisis is through creativity. (...)
>
> We could start a kind of dream about this material that has rich echoes in a number of literary figures: "The Picture of Dorian Gray," Fellini's "Casanova"... (...) recalling the film "Indiana Jones in Search of the Lost Arc", we could speak of "Indiana Jones in search of the dreamy thoughts of the waking", which is the emotional reality which befalls the patient and the analyst during the session, starting from the internal world and the story.

Here is another example of how Ferro uses the meta-theory of hologram, on which his analysis is based, as a metaphor for the clinical explanation of what happens in the session.

Note how, from the meta-theoretical complex perspective of the hologram he proposes a simple metaphor expressing the essence of the session.

> "As far as here and now is concerned, can you remember the mathematical theory of fractals? According to this theory the same structure continuously repeats itself, on large scale, on middle scale and on a small scale. Therefore what happens during the session reproduces on a micrometric scale what can happen in the real and in the internal life. It is enough for me to look through the microscope of the session, considering the relative nature of all the stories in which the patient must be discredited in what they say.
>
> For example if a patient tells me that, as a child, his aunt used to shout in his ears, I do not interpret that as something happening during the session . Nevertheless I try to speak to him in a way that he does not feel me like his aunt shouting in his ears."

MATTERS OF STYLE

Main clauses seem to take a picture of the account and suggest that the clinical event occurs in real time; they are "cut, dry, sober" in a perfect coupè style (Frescaroli 1996).

Clauses are linked through *asyndetons* and very few conjunctions are used.

Ferro insists on a pressing rhythm achieved via a rapid and hinted isomorphism between emotions felt during a session and those described in the text, obtained through short clauses and very few subordinate clauses.

There is neither abundant description nor oratory style, the use of complements is minimal, and so is that of coordinate, subordinate and non-defining relative clauses.

Simple appositions - constructions in which a noun is placed with another as an explanatory equivalent- and opposite subordinate clauses are nearly unknown.

Sentences are in the style of contemporary narrative and their directness recalls masterpieces by Joyce, Kafka and more recently those by Buzzati and Margherite Duras. Description always prevails against mere argument and the complexity of emotional entanglement is expressed though the acuity of a furtive glance.

Of the reader is requested minimum effort thanks to an immediate immersion in the story with no preambles and introductory frills.

The reader is immediately on the scene, the quantum jump is absolute, instantaneous, and a sign of a discontinuity hinting at multiple possible openings of meaning, at multiple possible worlds as modal logic would indicate.

It is even quicker thanks to a present-tense narration cadenced on full stops.

Narrative rhythm is therefore more and more pressing, and may follow the emotional alternation of the session, characters can be evanescent and faithful like the emotions during the session.

Efficiency is assisted by the way of linking the periods of clauses. As is known from period analysis, *connection* is affected by the astuteness of allowing prepositions to flow smoothly. Here this is obtained alternating the direct and indirect forms of the discourse. Thus the subject of the indirect form becomes that of the direct form and vice versa.

That is obtained by alternating direct and indirect speech hence the subject of the direct form becomes the subject of the indirect form and vice versa.

The author thus achieves a bright and breezy effect describing the crossing of the projective identifications circulating in the emotional field and stimulates the multiple possible interpretations according to the respective points of view.

Furthermore clauses are originally linked using "emotional pictograms", resulting from the analyst's *reveries* and their derivatives of equal co-textual worth, the patient's oneiric *flashes* and night dreams.

From a linguistic point of view, these *psychic objects* consist of metaphors of evocative images, which, like analogical waves, generate continuous narrative developments.

Consider for example how the character taken to a session progressively becomes the center of the whole narrative; they become a character-metaphor, a real descriptive vertex, from which the narrative journey starts, fulcrum of the clinical work.

The metaphors Ferro uses are characteristic, similar to a cinema zoom, shameless and efficient, on emotions at play.

This play develops through the patient's free verbal associations, in a typically metaphorical and relational shape, like each of their messages, according to the psychoanalytic perspective, from which presentation of significantly evocative ones is facilitated.

Moreover Ferro abundantly uses some literary devices, clearly leading to the production of further images and metaphors.

In particular similes, metaphors, metonyms and synecdoches are frequent, which combined with a generally simple and bare style, stand out and propose themselves as fundamental elements of the plot.

The frequent use of these literary devices seems to have a precise didactic aim, since the reader-therapist is increasingly stimulated to generate their own meaningful images in order to understand the text and thus is attracted to the heart of the co-re-creation of the clinical case.

The detail of narration is often the element originating the pregnancy of the metaphorical image and it is then as often used and transformed through paraphrase, improving the elegance of the clinical reading.

We find an example of this use of the paraphrase in a supervision comment formulated by Ferro in

In Supervision in Psychoanalysis: The Sao Paolo Seminars: "Remembering the film *Raider of the Lost Ark,* Indiana Jones , in the search of the lost ark , could become Indiana Jones , in the search of the waking dream thought, which is the emotional truth of what happens during the session, starting from the internal world and the story".

Note how in this passage an image metaphor is used as a paraphrase which enriches and reconfigures the discourse, promoting new paths of meaning (Ferro 1992, 1996, 200).

GLOSSARY

Autopoiesis A network of processes which generate the same components by which they are themselves generated a system thus made up is able to resist perturbations without losing its own inner organization and can evolve into even more complex organizations. Transformations which take place according to this logic are typical of highly complex natural and cognitive systems. The analytical field also is an autopoietic system, whose constitutive relationships are produced by the intertwining of the reciprocal projections in the field.

Dissipative structure (open system) An open system able to react to environmental perturbations dissipating disorder and maintaining its own functional organization.

Emerging quality A result of a non-linear generative process.

Entropy A measure of the disorder or randomness in a domain of phenomena. It is dissipated by open systems thus contributing to an organized evolution.

Hologram A complex organization where the whole and the part are in a reciprocally inclusive relationship: The whole is in the part which is in the whole: The whole and the part reciprocally retroact defining a unitary transformational field.

Non-linearity The logic of complex systems, evolving according to a deterministic principle with effects which are disproportionate to the causes and **quite** unpredictable

Metaphor A communicative device, which, through interpretation, is suitable to expand meaning and generate further horizons of comprehensibility.

Principle of competition A fundamental principle of systems evolving in a complex way through the dialectic process of order/disorder.

Projective identification A specific constitutive relationship of the autopoietic relation of the analytic field; it can be also defined as the relationship, coupling intrapsychic systems in an autopoietic unity at a higher level of the field which comprise them.

Systemic coupling A process through which two autopoietic unities become one by the intersection of their respective constitutive relationships. Two intrapsychic systems can become one, defining an analytical field which concerns and restructures both the components according new, more complex procedures. In this case the coupling takes place though the reciprocal projective identifications between patient and analyst.

Bibliography

Algini M.L. (2002), Non dire mai "lasciarci" in *Psicoterapia Psicoanalitica*, Anno IX, N.2, Luglio/Dicembre 2002, p. 39.

Arrigoni M.P., Barbieri G. (1998), Narrazione e psicoanalisi, un approccio semiologico, *Raffaello Cortina Editore*, Milano, 1998, p. 134, 138, 140.

Atlan H. (1971), Sul rumore come principio di auto-organizzazione, in *Teorie dell 'evento a cura di Edgar Morin*, Bompiani, Milano, 1972, p. 47.

Aulagnier P. (1975), La violence de l'interprétation.' du pictogramme l 'énoncé, *PUF*, Parigi, 1975.

Bach S. (1985), *Perspectives on self and object, in Narcissistic and the Therapeutic Process*, New York, Jason Aronson.

Bachelard G. in *M0n'n Edgar* (1990), Introduzione al pensiera complesso, Sperling & Kupfer, *Piacenza*, 1993, p. 12.

Balestrieri A. (1998), Gli istinti nell'u0m0, etologia, psicolagia e psicopa - tologia, *La Garangola*, Padova, 1998, p. 32.

Baranger Willy e Madeleine, La situazione psicoanalitica come campo bipersonale, *Raffaello Cortina Editore*, Milano, 1990. Idem b, p. 141. Idem c, p. 155. Idem d, p. 162.

Bama A. (1992), Ermeneutica e psicoanalisi, *Rivista di Psicoanalisi*, 1992, 3, p. 755.

Beccarla G.L. (1994), *Dizionario di linguistica*, Einaudi, Torino, 1994.

Beck E. (1991), La metafora come mediatore tra pensiero semantico e pen- siero analogico, in Teorie della metafora, a cura di C. Cacciari, *Raffaello Cortina Editore*, 1991.

Benvenuto S. (2001), Gli amori di matematica e psicoanalisi in *Il Vaso di Pandora*, Vol. IX, N.3, 2001, p. 18.

Bezoari M., Fiamminghi A.M. (1995), F unzione analitica e funzione genitoriale: alcuni modelli a confronto, in *Rivista di Psicoanalisi*, 1995; XLI; 2, p. 225.

Billow R. (1981), Observing spontaneous metaphor in children in *Journal of Experimental Child Psychology*, 31.

Bion W. (1962), *Apprendere dall 'esperienza*, Armando, Roma, 1972, p. 94, 158, 159.

Bion W. (1962), op. cit., p. 127.

Bocchi G. Ceruti M. (1985), *La sfida della complessità*, Campi del sapere, Feltrinelli, Milano, 1992, p. 7.

Bodei R. (2000), Le logiche del delirio. Ragione, affetti, follia, *Editori Laterza*, Bari 2000, p. 36. '

Bohr N. (1927), Teoria dell 'atomo e conoscenza umana, tr.it. Boringhieri, Torino, 1961.

Bohr N. (1938), Filosofia naturale e culture umane, in *Teoria dell 'atomo e conoscenza umana, tr* It. Boringhieri, Torino, 1961.

Bollas C. (1987), *The Shadow of the Object: Psychoanalysis of the Unthought Known*; Columbia University Press, New York 1987, in Mitchell, Speranza e timore in psicoanalisi, Bollati Boringhieri, Torino, 1995.

Bordi S. (1989), La tecnica psicoanalitica storia e mutamenti, in *Rivista di Psicoanalisi*, 1989; 3, p. 551.

Bridgman P.W. (1927), *Logic of Modern Physics*. Bunge M., Casualita, Harvard University Press, Cambridge Mass. 1959.

Capra F. (1997), *La rete della vita*, Rizzoli, Milano, 1997, p. 250.

Cargnelutti E., Muratori A.M. (1988), Trasformazioni in psicoanalisi, in *Rivista di Psicoanalisi*, N.3, Borla, Roma.

Cargnelutti E., Muratori A.M. (1990), Trasformazioni cognitive in psiconalisi, *in Rivista di Psicoanalisi*, N.2, Borla, Roma.

Carnap R. (1939), Foundations of Logic and Mathematics, in *Encyclopedia of Unified Science*, 1939.

Cartesio R., *Cogitationes privatae.*

Ceruti M. (1989*), La danza che crea, evoluzione e cognizione nell 'epistemologia genetica*, Feltrinelli, Milano, 1989.

Ceruti M., Lo Verso G. (1998), Sfida della complessità e psicoterapia, in *Epistemologia e Psicoterapia*, Complessità e frontiere contemporanee a cura di M. Ceruti e G. Lo Verso, Raffaello Cortina, Milano, 1998.

Chandler M. (1988), *Doubt and developing theories of mind*, in Astington J.W. et. a.l. (eds) Developing Theories of Mind, Cambridge Univ. Press Cambridge.

Chianese D. (1988), La costruzione di una teoria. Einstein-Freud: un confronto tra modelli, in *Rivista di Psicoanalisi*, 1988; 3, p. 485.

Cimatti F. (2003), Coscienza e linguaggio, in *Rivista di Psicoanalisi*, N .1, 2003, Borla, Roma, p. 24.

Ciompi L. (1995), L'intervento di crisi, in *Il Vaso di Pandora*, V01. III, N. l.

Corrao F. (1986), Il concetto di campo come modello teorico in *Gruppo e Funzione Analitica*, VII°, 1, Gennaio/Aprile 1986.

Corrao F. (1989), Morfologia e trasformazioni dei modelli analitici, in *Rivista di Psicoanalisi*, N.3, Borla, Roma, p. 529-531.

Correale A. (1998), Relazioni tra campo ed empatia in *Emozione e interpretazione psicoanalisi del campo emotivo*, Bollati Boringhieri, Milano, 1997, p. 203.

Cremerius J. (1985), *Il mestiere dell 'analista*, Boringhieri, Torino, 1985, p. 218.

Denis Hill J. (1970), The Contribution of Psychoanalysis to Psychiatry, *Int. J. Psycho Anal.* 49, 1-12.

Di Maria F., Giannone F. (1998), Epistemologia e scientificitàdel qualitativo, in *Epistemologia e Psicoterapia*, Complessità e frontiere contemporanee, a cura di M. Ceruti e G. L0 Verso, Raffaello Cortina, Milano, 1998.

Di Nuovo S., Giannone F., Di Blasi M. (1998), Epistemologia della valutazione in psicoterapia, in *Epistemologia e Psicoterapia, Complessità e frontiere contemporanee*, a cura di M. Ceruti e G. L0 Verso, Raffaello Cortina, Milano, 1998.

Downing G. (1995), *Il corpo e la parola*, Astrolabio, Roma, 1995.

Eco U. (1984), *Semiotica e filosofia del linguaggio*, Einaudi, 1984.

Einstein A. (1921), *Geometric und Erfarhung*.

Einstein A. (1931), *Maxewell 's influence on the development of the conception of physical reality*.

Einstein A. (1933), *On the method of the theoretical physics*.

Einstein A. (1944), *Remarks on Bertrand Russel is theory of knowledge*.

Emde R.N. (1980), Toward a Psychoanalitic Theory of Affect, in S.I. Greenspan, G.H. Pollock (eds), *The Course of Life: Psychoanalitic Contributions Toward Understanding Personality Development*, vol. 1, Mental Heat Study Center, Washington.

Eysenck H. (1961), Usi e abusi della psicologia, tr. it., *Edizioni Universitaria*, Firenze, 1961, p. 210.

Fairbairn W.R.D. (1958), Natura e scopi del trattamento psicoanalitico in *Il piacere e l'oggetto: scritti* 1952-1963, Astro1abio, Roma, 1992, p. 156.

Feyerabend P.K. (1984), Scienza come arte, *Biblioteca di cultura modema*, Laterza, Bari, P. 25.

Feyerabend P.K. (1984), op.cit. p. 35.

Feyerabend P.K. (1984), op.cit. p. 50.

Feyerabend P.K. (1984), op.cit. p. 52.

Feyerabend P.K. (1984), op.cit. p. 124.

Feyerabend P.K. (1975), Contro il metodo, *abbozzo di una teoria anarchica della concoscenza*; Feltrinelli, Milano, 1984, p. 155.

Feyerabend P.K. (1975), op.cit. p. 240.

Ferreri D. (1992), Nota introduttiva all 'edizione italiana, in *Il limite primigenio dell 'esperienza*, Thomas H. Ogden, Astrolabio, Roma, p. 18.

Ferri P., *L'età digitale e le sue epistemologie*.

Ferro A., Nella stanza di analisi, *Raffaello Cortina*, Milano, 1996.

Ferro A., La tecnica nella psicoanalisi infantile, *Raffaello Cortina*, Milano, 1992.

Ferro A., Teoria e tecnica nella supervisione psicoanalitica, *Raffaello Cortina*, Milano, 2000.

Ferruta A., Galli T. (1992), Ripetizione e sopresa in analisi, *Rivista di Psicoanalisi*, 1992, 3, p. 595.

Flavell J.H. (1988), The development of children's knowledge about the mind: from cognitive connections to mental representations in Astington JW et al. (eds), *Developing Theories of Mind*, Cambridge Univ. press Cambridge.

Fonagy P. (1991), Thinking about thinking: some clinical and theoretical considerations in the treatment of a borderline patient, *Int. Journal Psychoanal.* 72 p. 639-656.

Fossi G. (1991), La trasformazione degli affetti, in *Rivista di Psicoanalisi*, N.2, 1991, p. 389.

Fraiberg A. (1983), Pathological defenses in infancy in *Psychoanalytic Quarterly*, 60.

Frank P. (1958), Einstein, Mach e il positivismo logico, in *A. Einstein Scienziato e filosofo*, Ed. Scientifiche Einaudi, 1958, P. 220.

Frescaroli A. (1986), *Saper scrivere bene oggi*, De Vecchi Editore, Milano, 1996.

Freud S. (1892-95), Studi sull'i.§te_ria, in Sigmund Freud, *Opere*, Bollati Boringhieri, Torino, 1989, Vol. 1.

Freud S. (1895), A proposito di una critica della "nevrosi d 'ang0scia", in Sigmund Freud, *Opere*, Bollati Boringhieri, Torino, 1989, Vol. 2.

Freud S. (1895), Progetto di una psicologia, in Sigmund Freud, *Opere*, Bollati Boringhieri, Torino, 1989, Vol. 2, p. 201.

Freud S. (1895), Progetto di una psicologia, in Sigmund Freud, *Opere*, Bollati Boringhieri, Torino, 1989, Vol. 2.

Freud S. (1895), L'ereditarietà e l'etiologia delle nevrosi, in Sigmund Freud, *Opere*, Bollati Boringhieri, Torino, 1989, Vol. 2, p. 293.

Freud S. (1895), Nuove osservazioni sulle neuropsicosi da difesa, in *Sigmund Freud, Opere*, Bollati Boringhieri, Torino, 1989, V01. 2.

Freud S. (1896), L'ereditariet& e l'eti0l0gia delle nevrosi, in *Sigmund Freud, Opere*, Bollati Boringhieri, Torino, 1989, Vol. 2.

Freud S. (1896), Nuove osservazioni sui meccanismi di difesa, in *Sigmund Freud, Opere*, Bollati Boringhieri, Torino, 1989, V01. 2.

Freud S. (1899), L'interpretazione dei sogni, in *Sigmund Freud, Opere*, Bollati Boringhieri, Torino, 1989, Vol. 2.

Freud S. (1901), Frammento di un caso di isteria, in *Sigmund Freud, Opere*, Bollati Boringhieri, Torino, 1989, Vol. 2.

Freud S. (1905), Il motto di spirito e la sua relazione con l'inc0nsci0, in *Sigmund Freud, Opere*, Bollati Boringhieri, Torino, 1989, Vol. 5.

Freud S. (1905b), 0p.cit. p. 204.

Freud S. (1907), Il poeta e la fantasia, in *Sigmund Freud, Opere*, Bollati Boringhieri, Torino, 1989, Vol. 5, p. 383.

Freud S. (1908), Fantasie isteriche e loro relazione can la bisessualità, in *Sigmund Freud, Opere*, Bollati Boringhieri, Torino, 1989, Vol. 5, p. 390.

Freud S. (1908), Osservazioni generali sull'attacco isterico, in *Sigmund Freud, Opere*, Bollati Boringhieri, Torino, 1989, V01. 5, p. 441.

Freud S. (1911), Precisazioni sui due principi dell 'accadere psichico, in *Sigmund Freud, Opere*, Bollati Boringhieri, Torino, 1989, Vol. 6, p. 458.

Freud S. (1912), Totem e tabii, in *Sigmund Freud, Opere*, Bollati Boringhieri, Torino, 1989, V01. 7.

Freud S. (1914), Ricordare, ripetere e rielaborare, in *Sigmund Freud, Opere*, Bollati Boringhieri, Torino, 1989, Vol. 7.

Freud S. (l914b), Caso clinico dell'u0m0 dei lupi, in *Sigmund Freud, Opere*, Bollati Boringhieri, Torino, 1989, Vol. 7, pagg. 559-560.

Freud S. (1915), Pulsioni e loro destini, in *Sigmund Freud, Opere*, Bollati Boringhieri, Torino, 1989, V01. 8.

Freud S. (1915), L'inc0nsci0, in *Sigmund Freud, Opere*, Bollati Boringhieri, Torino, 1989, Vol. 8, p. 74.

Freud S. (1915), Introduzione alla psicoanalisi (I915-1917), in *Sigmund Freud, Opere*, Bollati Boringhieri, Torino, 1989, V01. 8, p. 526.

Freud S. (1918), Una nevrosi infantile, materiale aggiuntivo dai tempi remoti. Soluzione, (1918), in *Sigmund Freud, Opere*, Bollati Boringhieri, Torino, 1989, Vol. 7, pagg. 569-570.

Freud S. (1920), Al di Id del principio di piacere, in *Sigmund Freud, Opere*, Bollati Boringhieri, Torino, 1989, Vol. 9.

Freud S. (1921), Psicologia delle masse e analisi dell 'Io, (1921), in *Sigmund Freud, Opere*, Bollati Boiinghieri, Torino, 1989, Vol. 9, p. 280.

Freud S. (1922), L'Io e l 'Es, (1922), in *Sigmund Freud, Opere*, Bollati Boringhieri, Torino, 1989, V01. 9, p. 489.

Freud S. (1925), La negazione, in *Sigmund Freud, Opere*, Bollati Boringhieri, Torino, 1989, Vol. 10.

Freud S. (1926), Inibizione, sintomo e angoscia (1926), in *Sigmund Freud, Opere*, Bollati Boringhieri, Torino, 1989, V01. 10.

Freud S. (1929), Il disagio della civiltà, in *Sigmund Freud, Opere*, Bollati Boringhieri, Torino, 1989, V01. 10, pagg. 621-622.

Freud S. (1932), Introduzione alla psicoanalisi nuova serie di lezioni (1932), in *Sigmund Freud, Opere*, Bollati Boringhieri, Torino, 1989, Vol. 11, p. 263.

Freud S. (1937), Analisi terminabile e interminabile, in *Sigmund Freud, Opere*, Bollati Boringhieri, Torino, 1989, V01. 523.

Freud S. (1938), L'uomo Mose e la religione monoteistica, in *Sigmund Freud, Opere*, Bollati Boringhieri, Torino, 1989, V01. 11.

Gadamer H.G. (1960), *Verita e metodo,* Bompiani, Milano, 1983.

Gaddini E. (1982), Early defensive phantasies and the psychoanalytic process in *A. Limentani* (a cura di), A Psychoanalytic Theory of Infantile experience: Conceptual and Clinical Reflections, Routledge, London, 1992.

Galenson E. (1986), Some thoughts about infant psychopathology and aggressive development in *International Rev. Psychoanalysis*, 13.

Garella A., Napoletano F., Confini critici in psicoanalisi. La dinamica psi- chica in Freud e la teoria della complessità, in *Rivista diPsicoanalisi*, 1996; XLII; 2, p. 249, p. 256.

Gargani A. (1988), *Sguardo e destino*, Laterza, Roma-Bari.

Gauld A., Shotter J. (1983), *L'azione umana, tr. it.*, Città Nuova, Roma, 1983.

Gesué A., Sabucco G. (2000), Etica nelle istituzioni, in *Argonauti*, n. 84, Marzo 2000, CIS Editore, Gessate Milano, p. 47.

Glymour C. (1974), Freud, Kepler and the Clinical Evidence, in *Wollheim R.* (a cura di), Freud, Anchor Books, New York, 1974.

Greemberg J .R., Mitchell S.A. (1983), *Le relazioni oggettuali nella teoria psicoanalitica,* Bologna, Il Mulino, 1986, p. 71.

Green A. (1975), The analyst, symbolisation and absence in the analytic setting (On changes in analytic practice and analytic experience), *International Journal of Psycho-Analysis*, 56.

Grimberg L., Grimberg R. (1975), *Identidad y cambio*, Ediciones Kargieman, Buenos Aires.

Groumbaum A. (1984), I fondamenti della psicoanalisi, Il Saggiatore, *Arnoldo Mondatori editore*, Milano, p. 15.

Groumbaum A. (1984), op. cit., p. 81.

Groumbaum A. (1984), op. cit., p. 66.

Guiraud P. (1950), *Psychiatrie générale*, Le Francois, Paris.

Guntrip H. (1971), *Psychoanalytic theory, therapy and the self*, New York, Basic Books.

Halbwachs F. (1971), Causalità lineare e causalità circolare in fisica, in *Leteorie della causalità*, M. Bunge, F, Halbwachs, T.S. Kuhn, J. Piaget, L. Rosenfield, Giulio Einaudi Editore, Milano, 1974, p. 41.

Halbwachs, op. cit. p. 72.

Hartmann H. (1950), Considerazioni sulla teoria psicoanalitica dell 'I0, in *Saggi sulla psicologia dell 'Io*, Boringhieri, Torino, 1976.

Hartmann H., *Psicologia dell 'Io e problema dell 'adattamento*, Boringhieri, Torino, 1973.

Hatwood M. (2002), *Negoziando con le ombre*, Ponte alle Grazie, Milano, 2002.

Hénon M., Heiles C. (1964), The Applicability of the Third Integral of Motion: Some Numerical Experiments, in *Astronomical Joumal*, n.69.

Huebuer K. (1969), Was zeight kepler's "astronomia nova" der modernen wissenschafstheorie?, in *Philosophia naturalis*, Vol. 11, 1969, p. 267.

Husserl E. (1959), *La crisi delle scienze europee e la fenomenologia trascendentale*, Milano, Il Sagittario, 1961.

Jacobson E. (1954), Self and the object world in *The Psychoanalitic Study of the Child*, in International Universities Press, New York, 9.

Jones E. (1959), Editorial Preface to S. Freud, *Collected Papers*, Vol. 1, Basic Books, New York, 1959.

Kauffrnan S. (1993), The Origins of Order, Oxford University Press, New York, 1993, in Capra F. (1997), *La rete della vita*, Rizzoli, Milano, 1997, p. 250.

Kemberg O. (1976), Normal Clinical Psychoanalysis and pathological development in *Object Relations Theory*.

Kemberg O. (2000), Aggressività, disturbi della personalità e perversioni, *Raffaello Cortina Editore*, Milano, 2000.

Knyazeva E.N., Kurdymov S.P., *Synergetics at the Crossroads of the Eastern and the Western Cultures*, Keldish Institute of Applied Mathematics dell'Accademia delle scienze di Russia, 1994.

Kohut H. (1977), *La guarigione del Sè*, Boringhieri, Torino, 1980, p. 88.

Kohut H. (1978), *La ricerca del Sè, Boringhieri*, Torino, 1990, pagg. 79, 89, 92, 94, 140.

Kohut H. (1982), Introspection, empathy and the semicircle of mental health in *International Journal of Psychoanalysis*, 63, 1982.

Kuhn T.S. (1962), *La struttura delle rivoluzioni scientifiche*, tr.it. Einaudi, Torino, 1969.

Kuhn T.S. (1962), op. cit., p. 10.

Kuhn T.S. (1971), La nozione di causalità nello sviluppo della fisica, *in Le teorie della causalità*, M. Bunge, F. Halbwachs, T.S. Kuhn, J . Piaget, L. Rosenfeld, Giulio Einaudi Editori, Milano, 1974.

Kuhn T. (1977), *La tensione essenziale, cambiamenti e continuità nella scienza*, Einaudi, Torino, 1985, p. 324.

Langs R. (1976), The Bipersonal Field, Jason Aronson, New York. Langton C., Artificial life, in Nadel L., Stein D. (a cura di), 1991, Lectures in Complex Systems, *SFI Studies in the Sciences of Complexity*, Lect. Vol. IV, Addison-Wesley, 1992.

Laplanche J. e Pontalis J-B (1967), Enciclopedia della psicoanalisi, *Economia Laterza*, Bari, 1993.Laszlo E. (1992), Il pericolo e l'opportunità, presentazione di Ilya Prigogine, Sperling & Kupfer, Milano, 1992, p. 2.

Lenti G. (1998), Al di la del principio di entropia: per una teoria evolutiva della relazione terapeutica, in *Il vaso di Pandora*, Vol. VI, N.1.

Lenzen V.F. (1958), Teoria della conoscenza di Einstein, in Albert Einstein scienziato e filosofo, *Edizioni Scientifiche Einaudi*, Torino, 1958.

Lichtenberg J.D., Lachmann F.M., Fosshage J., *Self and Motivation Systems: toward a Theory of Technique*, Analytic Press, Hillsdale, 1992.

Lorenz E. (1963), Deterministics Nonperiodic Flow, in *Journal of the Atmospheric Sciences* N. 20.

Lukacs in Morin Edgar (1990), *Introduzione al pensiero complesso*, Sperling & Kupfer, Piacenza, 1993, p. 12.

Lupasco S. (1962), L'Energie et la Matiére vivante. *Antagonisme constructeur et logique de l 'he'te'rogene*, Julliard, Paris, riedizione del 1974.
MacKinnon D.W. and Dukes W.F. (1964), Repression, in Postman L. (ed), *Psychology in the Making*, New York, Knopf, 1964, p. 703.
Mahler M., Pine F., Bergman A. (1975), *The Psychological Birth of the Human Infant: Symbiosis and Individuation*, Basic Books, New York, 1975.
Mancia M. (1988), Una nota su Daniel Stem e il mondo interpersonale del bambino, in *Rivista di Psicoanalisi*, 1998; N.1, Borla, Roma.
Manfredi S., Ferro A. (1990), Introduzione all 'edizione italiana, in *La situazione psicoanalitica come campo bipersonale*, Raffaello Cortina, Milano, 1990, p. X.
Margulis L. (1995), Gaia is a Tough Bitch, in *John Brokman, The Third Culture*, Simon and Schuster, New York, 1995, p. 346.
Marti E. (1979), *La pensée analogique chez l 'enfant de 2 a 7 ans*, Tesi di dottorato, Universita di Ginevra.
Maruyana M. (1976), Toward Cultural Symbiosis, in Jantsch E., Waddington C.H. (a cura di), *Evolution and Consciousness: Human Systems in Transition*, Addison-Wesley, reading (Mass.).
Maturana H., Varela F. (1980), *Autopoiesi e cognizione*, Saggi Marsilio, Venezia, 1992, p. 160, 170 e 201.
Meandri R. (1989), Note sugli stili interpretativi in psicoanalisi, in *Rivista di Psicoanalisi*, 1989, N.1, Borla, Roma, p. 13.
Meltzer D. (1993), Claustrum, Uno studio dei fenomeni claustrofobici, *Raffaello Cortina editore*, Milano, 1993, p. 50.
Meltzer D. (1973), Stati sessuali della mente, *Armando Editore*, 1983.
Meltzer D. (1992), Claustrum, *Raffaello Cortina Editore*, Milano, 1997.
Meltzer D. (1998), *Lavorando con Meltzer*, Armando, Roma.
Magone P. (1995), *Terapia psicoanalitica*, Seminari, Franco Angeli, Milano, Italia, 1995.
Mitchell S. (1993), *Speranza e timore in psicoanalisi*, Bollati Boringhieri, Torino, 1995, p. 71 e 113.
Modell A. (1968), *Object love and reality*, New York, International University Press, trad. it. Amore oggettuale e realtà, Torino, Boringhieri, 1978.
Monakov C., Morgue R. (1928), *Introduction a l 'étude de la neurologie et de la psychopatologie*. Integration et disintegration de la function, Accan, Paris, 1928.
Morin E. (1969), *Il vivo del soggetto*, Moretti e Vitali Bergamo, 1998, p. 75, 302, 317, e 366.
Morin E. (1980), *La vita della vita*, Feltrinelli, Milano, 1987, p. 34.
Morin E. (1985), *Le vie della complessitaà in La sfida della complessità*, Bocchi Ceruti, Feltrinelli, Milano, 1989, p. 74 e. 232.
Morin E. (1986), *La conoscenza della conoscenza*, Feltiinelli, Milano, 1989, p. 74 e 232.
Morin E. (1990), *Introduzione al pensiero complesso*, Sperling & Kupfer Editori, Milano, p. 74.
Morin E. (1990, op. cit. p. 57.
Morin E. (1991), *Le idee: habitat, vita organizzazione usi e costumi*, Feltrinelli, Milano, 1993, p. 212.
Morin E. (2000), *La testa ben fatta*, Raffaello Cortina Editore, Milano, 2000.
Morlotti R. (1999), Conoscenza *della metafora/metafora della conoscenza*, Comunicazione Congresso Redancia, 1999, Savona.

Morlotti R. (2002), Linguaggio, conoscenza ed invenzione: essere in ginoco, in *Il Vaso di Pandora*, Vol. IX, N.2, 2002, p. 86.

Moszkowski A. (1921), Einstein, 1921. Napolitano F. (1988), Un approccio coerentista a un problema metapsicologico, in *Rivista di Psicoanalisi*, N.3, Borla, Roma, p. 439.

Northrop (1958), La concezione della scienza di Einstein, in *A. Einstein scienziato e filosofo*, Ed. Scientifiche Einaudi, Torino, 1958.

Ogden T.h. (1991), *L'identificazione proiettiva e la tecnica psicoterapeuta*, Astrolabio, Ubaldini Editore, Roma, p. 18 e 46.

Ogden T.h. (1997), *Reverie e interpretazione*, Astrolabio, Ubaldini Editore, Roma, pp. 61-62.

Pedone R. (1995), Artficial life: verso una biologia sintetica, in *Caos, Rivista scientifica ambientale di Legambiente*, Numero 4 (2/95), Edizioni Ambiente, Milano, 1995.

Piaget J. (1967), Logique et connaissance scientifique, *Encyclopedic de la Pléiade*, Gallimard, Paris, p. 1244.

Rapaport D. (1957), A theoretical Analysis of the Superego Concept. Parisi M. (1989), Situazioni di impasse e regolazione reciproca, in *Rivista di Psicoanalisi* N.4, 1989, Borla, Roma, p. 927.

Prigogine I., Stengers I. (1993), *La nuova alleanza*, La metamorfosi della scienza, Piccola Biblioteca Einaudi.

Ricoeur P. (1983), *Tempo e racconto*, Jaca Book, Milano, 1986.

Rispoli L. (1998), Verso un modello multidimensionale, in *Epistemologia e psicoterapia a cura di,* M. Ceruti e G. Lo Verso, Raffaello Conina Editore, 1998, Milano, p. 178.

Ritvo B.L. (1990), Darwin e Freud, *il racconto di due scienze*, I1 Pensiero Scientifico Editore, Roma, 1992.

Rossi P.L. (1988), Problemi attuali dell 'osservazione madre-bambino in America, in *Rivista di Psicoanalisi* N.1, 1988, Borla, Roma, p. 195.

Rubinstein B.B. (1975), On the Role of Classificatory Processes in Mental Functioning, Aspects of a Psychoanalytic Theoretical Model, in *Psychoanalysis and Contemporary Science*, 3, 1975, pagg. 104-105.

Ruelle D., Takens F. (1980), Strange Attractors, in *Mathematical Intelligencer,* N.2. Reichenbach H. (1958), Il significato fIlosofico della teoria della relativita, in *A. Einstein scienziato e filosofo*, Ed. Scientifiche Einaudi, 1958, p. 243 e 246.

Russo L. (1990), Ipotesi sul campo analitico, in *Rivista di Psicoanalisi*, anno XXXIV, N.4, 1990, p. 1113.

Sarno M. (1990), L'archivio della tecnica: il resoconto psicoanalitico, in *Rivista di Psicoanalisi,* 1990, p. 55.

Shannon C.E. e Weaver W. (1959), *The Mathematical Theory of Communication*, University of Illinois Press, Urbana III.

Schafer R. (1976), *A new Language for Psychoanalysis,* Yale University Press, New Haven, 1976, p. 204-205.

Schafer R. (1979), On becoming an analyst of one persuasion or another, in *Contemporary Psychoanalysis*, 15, pagg. 345-368.

Schafer R. (1983), The Analytic Attitude, Basic Books, New York, 1983, in *Mitchell, Speranza e timore in psicoanalisi*, Bollati Boeringhieri, Torino, 1995.

Schafer R. (1992), *Retelling a life*, Basic Books, New York, 1992.

Seganti A. (1991), La formazione del senso di colpa nella memoria aflettiva, in *Rivista di Psicoanalisi*, Borla, Roma, N.l, 1991, pagg. 203-205.

Siciliani O. (1970), Energia psichica o informazione? Evoluzione storica e attuale revisione del concetto di "energia psichica" in *Riv. Psichitr.* 5.

Snyder J. (1979*), The spontaneous production of figurative language and work play in the grade school years,* Tesi di Ph.D., Boston University.

Spence D.P. (1993), Sotto la superficie analitica, La teoria della mente del l'analizzando tr. it. *In Psicoterapia*, Anno 1, N.1, 1995, p. 119.

Stafford Beer In *Prefazione di Autopoiesi e cognizione*, Humberto R. Maturana, Francisco J. Varala, Saggi Marsilio, Venezia, 1992, p. 119.

Stengers I. (1988*), Da una scienza all 'altra*, Concetti nomadi, Hopefulmonster, Firenze, 1988.

Stem D. (1987), *Il mondo interpersonale del bambino*, Bollati Boringhieri, Torino, 1987.

Stolorow R.D. Lachmann F. (1980), *Psychoanalysis of Developmental Arrests: Theory and Treatment*, International Universities Press, New York.

Thom R. (1972), Stabilité culturelle et Morphogénese. *Essai d 'une théorie génétique des modeles*, Ediscience, Paris; tr.it. Stabilita strutturale e morfogenesi, Einaudi, Torino, 1980.

Tibaldi G. (1983), *Corpo inconscio simbolo*, Edizioni Cortina, Milano, 1983.

Tobler Stefan (2000), La relazionalità come metodo del pensare? Spunti per un 'ermeneutica del patto d 'unità, in *Nuova Umanità*, Rivista Bimestrale di Cultura, Città Nuova Editrice, Roma, 2000, p. 825.

Ueda (1985), Random Phenomena Resulting from Nonlinearity in the System Described by Dufling 's Equation, in *International Journal of Non-linear Mechanism*, n.20.

Varela F. (1984), The Creative Circle: Sketches on the Natural History of Circularity, in P. Watzlawick (a cura di), *The invented Reality*, Norton, New York.

Vassallo Torrigiani M.G., Vassallo S. (2001), L'autOritrattO nella came: il realismo psicotico di Orlan in *Rivista di Psicoanalisi* N.4 2001, Borla, Roma. pagg. 744, 745.

Valtimo G. (1965), Introduzione, in *Verità e metodo di H.G. Gadamer*, 1960, Fabbri Editori, 1965.

White M.T., Weiner M.B. (1986), *La teoria e la pratica della psicologia del Se'*, Astrolabio, Roma, 1988.

Winfree A. (1974), Rotating Chemical Reactions, in *Scientific American*, 1974.

Winnicott D. (1948 a), Pediatria e psichiatria, in *Dalla Pediatria alla Psicoanalisi*, Martinelli, Firenze, 1995, p. 228.

Winnicott D. (1948 b), ibidem, p. 197.

Winnicott D. (1948 c), ibidem, p. 239-244.

Wolf E.S. (1988), *La cura del Sé*, elémenti clinici di Psicologia del Sé, Astrolabio, Roma, 1993.

Zingarelli, Ontico, *definizione tratta dallo* Zingarelli 2001, Vocabolario della lingua italiana, Zanichelli, Bologna.

About the Author

Gabriele Lenti was born in Alessandria, Italy, on 19th April 1963. He lives and works in Genova. He is the author of numerous publications on professional journals including: Area di gioco (Playing Area) Psicoterapia Psicoanalitica (Psychoanalytical Psychotherapy) Il vaso di Pandora (Pandora's Box) Ricerca psicoanalitica (Psychoanalytical Research), and he has written two books: Al di là del principio di entropia. Alcune considerazioni su psicoanalisi e complessità, (Beyond the principle of entropy. Some Considerations about Psychoanalysis and Complexity) Armando, Roma, 2005; Gli incerti percorsi della conoscenza, Redancia, Savona, 2007. (The uncertain paths of knowledge) - collective work.

He is working on a third volume Complessità per una psicoanalisi futura, (Complexity for a future psychoanalysis) which will be probably published by Armando Roma.

He has given dozens of lectures on psychology and several at scientific level at universities, Italian psychoanalytical centers.

He has recently taken part with a personal paper at the IFPS (International Federation Psychoanalytic Societies) International Congress in Mexico City.

He is a member of SIPRE (Società Italiana Psicoanalisi della Relazione/ Italian Society Relational Psychoanalysis) IFPS and IARPP.

Mailing Adress: Via Chiariamone 12 16158 Genova Italy
E-mail address: gabrielelenti@gmail.com
Cell Number: 0039 339 2943612
Phone numbers: 0039 010 6131904 / 0039 010 5701615

(Translation Maura Pasero; Proofreading Don Pilarz)

INDEX

A

Abraham, 79
abstraction, 4, 5, 9, 11, 12, 24, 31, 32, 37, 38, 50, 72, 89, 120, 124, 130, 131, 141
access, 136
acid, 57
acquired characteristics, 17, 18
ADA, 119, 120
adaptation, 28, 31
adaptive functioning, 106
adjustment, 1, 80
adulthood, 4
adults, 26, 30, 36
aesthetic, 2, 128, 129
affective experience, 27
Africa, 150
age, 31, 37, 132, 150, 153
agglutination, 111
aggregation, 58, 134
aggression, 25, 77, 152
aggressiveness, 27, 28, 29, 38, 152
algorithm, 79
alters, ix
ambivalence, 141
amplitude, 67
anal stage, 30
anger, 27, 28, 146
antagonism, 81, 82, 129
antidepressants, 77
anus, 114
anxiety, 36, 38, 78, 114
appetite, 71
Aristotelian logic, xii, 12, 13, 82
Aristotle, 9, 35, 130
arousal, 27
arrow of time, 130
articulation, 131
artificial intelligence, 63
aspiration, 79, 141
assimilation, 123
asymmetry, 88
atmosphere, 28, 29, 30, 46, 135, 149
atoms, 8, 49, 97
attachment, 25, 37
attachment theory, 25
automata, 64
autonomy, 56, 89
aversion, 1, 40
awareness, xiii, 2, 18, 31, 33

B

bacterium, 20
base, xii, 10, 11, 72, 76, 113, 124
behaviorism, 10
behaviors, xii, 26, 55, 63, 64, 79, 106, 130, 131
benefits, 146
benign, 28
bias, 26
biochemistry, 56, 101
biological sciences, 50
biological systems, 88
bison, 150
black hole, 82
blind spot, 91
body shape, 120
bonds, 31, 97, 99
bone, 135
boredom, 150
bounds, 75, 76, 79, 81, 116, 130
breastfeeding, 26
budding, 28
building blocks, 116

C

carbohydrates, 85, 87
caregiving, 114
Carnap, 9, 160
castration, 18
catalyst, 57
catastrophes, 68
categorization, 32, 131
cathexis, 27, 152
causal attribution, 48
causal relationship, 52
causality, 3, 5, 9, 12, 13, 35, 36, 37, 38, 45, 47, 56, 59, 89, 93, 96
causation, 36, 93
central nervous system, 4
chaos, xi, xii, 21, 38, 65, 75
chaotic behavior, 63
charm, 10
chemical, 19, 49, 55, 101
chemical reactions, 55
childhood, 4, 18, 25, 77, 115, 128, 132, 135, 148
children, 18, 36, 77, 113, 132, 159, 161
circumcision, 18
CIS, 163
city(s), 67, 169
civil servants, 146
civilization, 38
classes, 41
classical mechanics, 56, 59
classification, 39
claustrophobic, 113, 136
clinical application, 134
clinical presentation, 4, 149
clinical trials, 45
cloaca, 115
closure, 92
coding, xii, 131
cognition, 132
cognitive function, 31
cognitive models, 95
cognitive perspective, 32
cognitive process, 11, 124, 130, 144
cognitive system, 116, 133, 157
cognitive tool, xi
cognitivist psychology, xii
coherence, 35, 45, 82, 106
collaboration, 13, 16, 61, 76, 127
collusion, 93
color, 131, 145
combinatorics, 62, 100
common sense, 68, 76
communication, 2, 13, 23, 63, 72, 80, 89, 100, 112, 130, 141, 147, 148
community, 1, 26, 36, 145
competition, 23, 56, 58, 76, 96, 143, 157
complement, 19, 51, 65
complementarity, 52, 118
complex interactions, xii
complex organizations, 157
complexity, ix, xi, 1, 4, 5, 7, 10, 11, 12, 17, 20, 21, 27, 28, 29, 33, 40, 44, 45, 51, 56, 61, 62, 68, 69, 70, 73, 80, 82, 84, 85, 88, 95, 97, 99, 107, 109, 117, 120, 124, 139, 143, 145, 149, 154
comprehension, xiii, 1, 2, 25, 50, 56, 62, 88, 100, 132, 134, 141
compulsion, 13, 14, 93, 111
computation, 21, 59
computer, 39, 53, 65, 76, 89
computer research, 89
computer simulations, 53, 65
computing, 87
conception, xiii, 3, 45, 47, 50, 103, 110, 161
conceptualization, 24, 83
conciliation, 129
condensation, 41, 133
conditioning, 123
conductor, 65
configuration, 55
conflict, 4, 13, 29, 130, 135
confrontation, ix
congress, 59, 169
connective tissue, 117
connectivity, 131
conscious experiences, 92
consciousness, 13, 95, 129, 133
consensus, 47
consolidation, 135
constituents, 76, 81
construction, xiii, 7, 8, 18, 41, 84, 87, 105, 110, 133, 136
consulting, 99, 110, 113, 147, 151
containers, 32, 84
contradiction, 15, 65, 82, 97, 141, 147
convention, 119
convergence, 23
conviction, 35
cooperation, 76, 127
copper, 145
correlation, 124
cosmos, 51, 115
cost, 143
creativity, 45, 130, 137, 153
creep, 47
critical analysis, 8

criticism, 2, 24, 45, 84, 143
cultural heritage, xi
culture, 110, 160
customers, 40
cycles, 20, 67

D

Darwinian evolution, 19
Darwinism, 17, 18, 19, 20
data processing, 89
decentralization, 143
decoding, xii, 131, 147, 151
degradation, 50
delegates, 112
delirium, 134
democracy, 33
denial, 111
depth, 10
derivatives, 154
despair, 153
destiny, 7, 32, 53, 67
destruction, 29, 38, 51, 105
detectable, 67
detection, 100
determinism, 4, 5, 11, 13, 47, 50, 51, 55, 59, 93, 97
deviation, 11
dichotomy, 1
dignity, xiii, 40, 70, 127
discontinuity, 51, 58, 137, 154
disorder, 21, 50, 51, 52, 53, 54, 55, 57, 61, 62, 63, 65, 68, 75, 76, 81, 82, 101, 117, 119, 129, 143, 157
disposition, 115
dissipative structure, 54, 55, 57, 61, 63, 64, 81
dissipative structures, 54, 55, 61, 63, 64
distillation, 147
distortions, 5, 46, 59, 91, 109, 130
distribution, 21
divergence, 59
diversification, 56
diversity, 20
dizziness, 134
DNA, 20, 21
drawing, 67, 152
dream, 40, 42, 46, 110, 115, 134, 149, 152, 153, 155
dreaming, 78, 134
drive theory, 26, 30, 31
dualism, 4, 32
duality, 130
dynamic systems, 58

E

E.coli, 20
earthquakes, 21
ecology, 19
economic theory, 16
elaboration, 69, 71, 96, 133
electricity, 35, 36
elementary particle, 35, 51
elucidation, 96
emotion, 16, 134, 148, 151
emotional experience, 127
emotional state, 148
empathy, 30, 105, 106, 110, 164
energy, 4, 8, 16, 17, 20, 24, 27, 32, 36, 49, 50, 51, 57, 58, 61, 64, 69, 97
energy supply, 50
entropy, xiii, 8, 50, 53, 57, 61, 64, 81, 119, 169
environment(s), xiii, 15, 17, 19, 20, 26, 27, 28, 29, 30, 37, 54, 57, 58, 61, 87
environmental factors, 15, 30
environmental stimuli, 25
environmental variables, 57
enzymes, 55
episteme, 82
epistemology, 8, 31, 35, 69, 72, 73, 82, 84, 95, 100, 129, 139, 145
equilibrium, 17, 21, 28, 54, 55, 57, 58, 67, 69, 117, 141
etiology, 4, 5, 15, 17
Euclidean space, 68
evidence, 7, 13, 36, 44, 46, 51, 115, 145
evolution, xiii, 16, 17, 18, 19, 21, 24, 31, 49, 53, 54, 59, 63, 67, 73, 75, 81, 84, 87, 97, 98, 100, 112, 115, 124, 128, 131, 145, 150, 157
exclusion, 46, 52
expert systems, 63
external environment, 61
extinction, 29
extraction, xi

F

facilitators, 129
faith, 4
fantasy, 18, 28, 78, 91, 111, 112, 113, 123, 128, 129, 148, 151
fat, 135
fear, 38, 114, 136
feelings, 26, 27, 78, 83, 111, 114, 123, 131, 132, 133, 136, 143, 146, 149, 151
fidelity, 120

field theory, 73, 92, 101, 127
films, 150, 152
filters, 23
fixation, 4, 141
flexibility, xiii
flight, 110
flowers, 115
fluctuations, 21
fluid, 68
food, xii, 20, 109, 135, 146
force, 17, 21, 23, 141, 143
formation, 11, 36, 41, 42, 53, 55, 63, 79, 105, 148
foundations, 9, 19, 49, 64
fractures, 81
fragments, 132
France, 45, 95
free association, 134
freedom, xiii, 8, 63, 79, 96, 125, 136, 148
frequency distribution, 69
Freud, v, xiii, 3, 4, 5, 7, 8, 9, 10, 11, 12, 13, 14, 15, 16, 17, 18, 19, 21, 23, 24, 25, 26, 27, 29, 30, 31, 32, 36, 37, 38, 39, 41, 44, 45, 47, 48, 72, 82, 83, 93, 99, 109, 110, 111, 128, 129, 130, 160, 161, 162, 163, 164, 166
friction, 67
friendship, 135
functional architecture, 51
functional fixedness, 11
fusion, 135, 143

G

galaxies, 68
Galileo, 9
generativity, 100, 101, 124
genes, 19, 20
genetic factors, 15
genetic information, 64
genetics, 19, 80
genome, 20
geography, 114
geometry, 8, 68, 116
Gestalt, 76, 79, 92, 95
Gestalt theory, 76, 79, 95
gestures, 134
grants, xiii
gravitation, 35
gravitational force, 35, 95
greed, 135, 136
growth, xi, 30, 51
guilt, 38, 77

H

habitat, 165
harmony, 8, 149
healing, 78
health, 114, 130, 146
heat transfer, 49
heredity, 5, 15, 18, 19
hermeneutics, 2, 48
history, 1, 3, 18, 19, 20, 21, 40, 46, 84, 133
HIV, 114, 146
holism, 82
hologram, 117, 148, 153
homeostasis, 69
hominids, 20
hostility, 135
human, 2, 5, 16, 18, 20, 28, 32, 39, 45, 46, 47, 69, 72, 96
human sciences, 2, 45, 46, 69, 96
humanistic psychology, 30
husband, 77, 147
hybrid, xi
hypertrophy, xiii
hypothesis, v, 14, 15, 18, 21, 27, 32, 39, 47, 92, 96, 114, 125
hysteria, 4

I

ideal(s), 1, 9, 21, 28, 50, 67
idealization, ix, 27, 77, 107
identification, xii, 24, 25, 27, 72, 77, 84, 91, 101, 103, 109, 110, 111, 112, 113, 115, 120, 123, 124, 157
identity, 12, 19, 31, 88, 89, 97, 103, 106, 113, 146, 150
ideology, 40
illusion(s), 25, 26, 78, 114
image(s), xi, 9, 41, 105, 106, 115, 130, 131, 132, 133, 134, 135, 136, 137, 146, 152, 154, 155
immersion, 154
impulses, 29
incompatibility, 32
independence, 3, 30, 135, 136, 150
independent variable, 105
indeterminism, 13
individual personality, 133
individuals, 19, 23, 25, 29, 110
individuation, 26
infancy, 161
infection, 146
inferences, 40

infiniteness, 124
inheritance, 17, 18, 19, 20
initial state, 55
inner world, 5, 12, 24
insertion, 59
instinct, 4, 32
integration, xiii, 26, 27, 29, 30, 31, 71, 75, 123, 134
intelligence, 49, 64, 135
intentionality, 30
intercourse, 115
interdependence, 13
internalization, 25, 27, 103
interpretability, 42
interrelations, 33
intervention, xii, 92
introspection, 36, 46
investment, 26, 27, 28
iron, 150
islands, 68, 75
isolation, 38, 137
issues, 64
Italy, ix, 169

J

Japan, 68
justification, 40

K

kill, 150
kinetics, 97

L

labeling, 92
lactose, 20
Lamarckism, 17, 18
languages, xiii, 11, 39, 48, 88
latent content, 92
laws, 1, 3, 10, 17, 43, 57, 62, 81, 89, 131, 133
lead, xi, xiii, 11, 13, 40, 57, 64, 95, 100, 114, 141, 148
learning, xii, 31, 39, 71, 95, 103, 104, 124, 132
liberty, 77
libido, 4, 24, 25, 27, 28, 29
life experiences, 133
light, 2, 5, 15, 37, 84, 99, 116, 130
linear logic, xi, 39, 62, 72, 118
linear model, 69
linguistic rule, 131
lipids, 85, 87
locus, 67
logarithmic functions, 119
longevity, 25
love, 27, 147, 148, 165
loyalty, xii, 40

M

magnetic field(s), 97
magnetism, 35, 36
magnitude, 36
malaise, 110, 150
man, 16, 35, 40, 41, 45, 77, 114, 129, 153
manipulation, 133
mass, 36, 58, 65, 67
masterpieces, 154
mathematics, 8, 41, 65, 97
matrix, 2, 110
matter, xiv, 12, 19, 27, 29, 32, 35, 36, 41, 43, 45, 50, 51, 56, 68, 69, 73, 105, 110, 143, 144, 149
measurement, 109
mediation, 133
medical, 65, 72
memory, xii, 4, 27, 63, 110, 119, 130, 152
mental activity, 96
mental health, 25, 146, 164
mental illness, 15
mental representation, 28, 161
mental state, 100
Merleau-Ponty, 91
messages, xii, 154
metabolism, 20
metabolized, 110, 112, 134
metaphor, 5, 12, 32, 39, 41, 46, 73, 83, 99, 116, 130, 131, 132, 133, 134, 137, 145, 148, 149, 151, 153, 154, 155, 159
metapsychology, 7, 13, 26, 139
methodology, 46, 48, 144
Mexico, 169
micro-events, 68
microorganisms, 21
microscope, 153
mildew, 58
mind-body, 107
misunderstanding, 134, 141
models, ix, xi, xii, 7, 19, 24, 27, 29, 31, 32, 33, 36, 38, 53, 54, 59, 68, 69, 84, 88, 89, 97, 127, 128, 143, 146, 149
modern science, 95
modernity, 16
modifications, 54, 87, 97
mold, 129
molecules, 3, 19, 54, 58, 85, 101

monopoly, 33
moral judgment, 115
Moses, 18
motivation, xii, 77
multicellular organisms, 58
multidimensional, 67, 68
multiplication, ix
multiplier, 5
murder, 17, 18, 21
mutation(s), 19, 20, 21, 132
mutilation, 28

N

naming, 143
narcissism, 24, 28
natural evolution, 19
natural laws, 129
natural science(s), 1, 2, 3, 4, 16, 21, 45, 76
natural selection, 16, 19, 20, 21
neglect, 1, 150
negotiating, 129
nervous system, 95, 134
neural network(s), 64
neurologist, 77
neuroses, 18, 29
neutral, 29, 79, 139
Niels Bohr, 96
Nietzsche, 75
nightmares, 134
non-Euclidean geometry, 8
non-linear equations, 59, 100
nonlinear logic, xi, 69
nostalgia, 5
nucleus, 18, 106, 110

O

objectification, 1, 46
objectivity, 2, 37, 39
oculomotor, 30
Oedipus complex, 18, 20
openness, xi, 15, 128, 137, 148
opportunities, 114
optimism, 10, 11
organ(s), 13, 17, 28, 75, 103, 113, 131, 134
organism, xii, 17, 18, 20, 62, 87, 97, 105
organize, 7, 16, 21, 27, 51, 54, 63, 81, 97, 101, 145
oscillation, 151
oxidation, 57
oxygen, 28, 37

P

pain, 17, 27, 39, 77
pairing, 110
paleoencephalon, xii
panic attack, 146
paralysis, 78
paranoia, 4
parents, 18, 28, 77, 78, 152
particle physics, 69
pathogenesis, 5, 15
pathology, 4, 17, 29, 32, 92, 96, 111, 130, 147
peace, 23, 146
penis, 147
perception, xii, 31, 33, 37, 38, 91, 105, 106, 131, 132
permit, xiii, 21, 32, 59, 64, 69, 92, 106, 121, 135
personal choice, 115
personal communication, 40
personality, xi, 12, 28, 109, 110, 111
persuasion, 166
phenomenology, 29, 91, 113
physical laws, 36
physical phenomena, 3, 59, 110
physical sciences, 36, 47, 59
physicalism, 48
physics, 3, 5, 7, 8, 10, 11, 12, 17, 21, 69, 96, 161
plasmodium, 58
Plato, 2, 8, 43, 131
plausibility, 15
playing, 152
pleasure, xii, 17, 24, 25, 27, 77, 128, 129, 130
pleasure principle, 17, 129, 130
positivism, 3, 5, 7, 10, 12, 14, 43
potassium, 57
predicate, 131
predictability, 37, 100
pregnancy, 155
prejudice, 2
preparation, 16
President, 109
principles, xii, 3, 4, 7, 8, 9, 12, 13, 33, 40, 45, 49, 51, 61, 64, 87, 97, 117, 129, 130, 131
probability, 41, 53, 55, 65, 100
producers, xii, 101, 117
project, 4, 29, 123
promoter, 97
propagation, 58
protection, 1, 78, 136
proteins, 85, 87
prototype, 36
psychiatrist, 16
psychic energy, 36
psychic process, 4, 71

psychoanalysis, ix, xi, xiii, 3, 5, 16, 18, 19, 23, 36, 39, 41, 45, 46, 47, 69, 73, 84, 91, 95, 96, 116, 127, 130, 139, 143, 145, 169
psychoanalytic theories, 25
psychological connections, 9
psychological phenomena, 96
psychological problems, 77
psychological processes, 123
psychologist, 44
psychology, ix, xii, 3, 4, 7, 15, 16, 17, 18, 23, 28, 29, 31, 56, 59, 169
psychopathology, 23, 24, 29, 114, 163
psychoses, 105
psychosis, 28
psychosomatic, 106
psychotherapy, xi, 77
punctuated equilibrium, 20
punishment, 38

Q

quantification, 3, 71, 80
quantum jumps, 8
quantum mechanics, 41, 47
questioning, 136

R

radiation, 8
rationality, 37, 40
reactants, 55
reactions, 15, 29, 55, 75, 77, 148
reading, ix, 143, 155, 165
real time, 153
realism, 9, 84
reality, xi, 2, 3, 8, 9, 10, 11, 14, 17, 18, 20, 24, 25, 26, 30, 33, 36, 37, 38, 39, 40, 41, 48, 53, 56, 78, 82, 83, 84, 91, 95, 98, 99, 100, 106, 107, 129, 130, 133, 136, 141, 151, 153, 161, 165
reasoning, 1, 2, 4, 9, 10, 13, 20, 32, 33, 40, 46, 52, 95, 131, 149
recalling, 93, 153
receptacle, 115
reciprocity, 101
recognition, 2, 38, 130
recombination, 19, 125
reconstruction, 148
recovery, 93
recurrence, 130
reductionism, 5, 76, 82
reference frame, 71
reflexes, 27
regenerate, 123
regeneration, 87, 131
regression, 92, 130
reinforcement, xii
rejection, 109
relapses, 69
relational constellations, xi
relational model, 32
relativity, 1, 10, 39, 72
relaxation, 25
relevance, 48
reliability, 26, 37
relief, 26
René Descartes, 137
repression, 4, 13, 44, 80, 129, 148
reproduction, 80
requirements, 151
researchers, xiv, 2, 36, 38, 68, 101
residuals, 133, 134
resistance, 12, 14, 91
resources, 40
response, 25, 28, 31, 65
restitution, 123, 151
restoration, 29, 53, 78
restrictions, 79
restructuring, 124
rewards, 19
rhetoric, ix
rhythm, 16, 145, 150, 153, 154
risk(s), 1, 72, 141, 146, 149
RNA, 20
role-playing, 99
root, 19, 120
routines, 81
rules, 4, 9, 11, 35, 37, 40, 41, 43, 67, 79, 80, 87, 88, 89, 91, 97, 99
Russia, 164

S

safety, 78, 147, 152
saturation, 130
school, 37, 45, 77, 132, 151, 167
science, ix, 1, 3, 5, 9, 10, 11, 17, 21, 33, 40, 41, 45, 46, 49, 50, 55, 62, 82, 84, 95, 96, 130
scientific knowledge, 9, 50
scientific progress, 40
scientific theory, 7, 11
scientific validity, 44
scope, 72
scripts, 147
sediments, 15, 87
self-observation, 91

self-organization, 20, 56, 64
self-regulation, 37
self-reproduction, 87
semicircle, 164
semiotics, 2
sensation(s), 9, 36, 133, 134, 148
senses, 41
sensitivity, 1
sensory experience, 27, 124, 131, 133
sex, 147
sexual abuse, 5
sexual development, 92
sexual experiences, 4
sexual pathology, 113
sexuality, 77, 114
SFI, 164
shame, 28, 128
shape, 27, 29, 31, 68, 69, 91, 106, 113, 115, 120, 127, 128, 147, 152, 154
showing, 128, 134
signals, xii, 28, 31, 150, 151
signs, 9, 88, 135, 141
simulation(s), 64, 65
skin, 136
smoking, 146
social organization, 51
social psychology, 110
social relations, 30
society, 23, 117
Socrates, 43
solution, 13, 64, 79, 95, 130, 132
space-time, 96
species, 16, 19, 20
specter, 146
speculation, 8, 99, 148
speech, 154
spin, 149
stability, 19, 20, 61, 62, 63, 68, 72, 75, 76, 81, 89, 101, 115, 129
state(s), xiii, 4, 8, 9, 12, 21, 25, 27, 39, 45, 48, 49, 50, 52, 55, 57, 58, 63, 67, 68, 69, 70, 72, 78, 81, 93, 95, 100, 103, 104, 116, 119, 120, 125, 128, 129, 133, 134, 136, 148, 151
sterile, 130
stimulation, 26, 131, 147
stimulus, 31, 55, 131
stomach, 134
storage, 87
stress, 127
structural changes, 88
structure, 1, 5, 8, 10, 12, 20, 24, 27, 28, 43, 69, 70, 78, 79, 83, 85, 91, 93, 97, 100, 101, 110, 127, 129, 153, 157
structuring, 28, 31, 82, 93, 97, 107, 111, 116, 121, 130, 145
style, ix, xii, 12, 40, 145, 153, 154
subjectivity, 2, 4, 27, 32, 40
substitutes, 33, 128
succession, 53
superconductor, 65
superego, 38
supervision, 134, 155
survival, 16, 56, 69
suspensions, 97
Switzerland, 69
symbiosis, 100
symbolic activity, xii
symbolism, 15
symmetry, xii, 63, 65, 76
symptoms, 2, 4, 16, 19, 24, 36, 65, 106, 128, 129, 130
synthesis, 17, 19

T

target, 64, 123
techniques, 29, 73
teleological, 19
temperature, 50, 53, 55
tension(s), 11, 17, 28, 30, 38, 123, 128, 141, 142
testing, ix
theoretical approaches, 10, 23, 30
therapeutic process, 75, 112
therapeutic relationship, 30
therapist, ix, 29, 92, 100, 106, 111, 124, 143, 154
therapy, 4, 47, 77, 100, 106, 115, 129, 135, 146, 163
thermal energy, 50
thermodynamic equilibrium, 50
thermodynamic properties, 50
thermodynamics, 17, 36, 49, 50, 51, 52, 53, 57, 58, 64, 76, 81
thoughts, 5, 33, 41, 42, 77, 105, 110, 115, 121, 124, 134, 135, 143, 147, 148, 153, 163
threats, 18
throws, 13
topology, 113
torture, 77
traditions, 40
traits, 20
trajectory, 67, 68, 136
tranquilizers, 77
transduction, 71
transference, 26, 78, 92, 93, 133, 147, 149, 151
transformation(s), xi, xii, xiii, 11, 31, 48, 49, 50, 68, 69, 71, 73, 75, 81, 87, 88, 92, 95, 97, 99, 100,

101, 110, 111, 116, 119, 120, 123, 127, 129, 130, 139, 141, 143, 147, 148, 149, 152
transgression, 133
translation, 82, 132, 147
transmission, 15, 101
transverse section, 69
trauma, 4, 5, 33, 47, 93
traumatic wound, xii
treatment, 24, 29, 77, 92, 105, 111, 128, 134, 151, 161
trial, 9
turbulence, 57, 58, 61, 62, 63, 65, 68, 69, 75, 129

U

unconscious fantasies, xi, 93, 99, 100, 114, 118, 128, 151
universe, 5, 10, 49, 51, 56, 58, 84
universities, 19, 169
urban, 21

V

vacuum, 30, 111
validation, 45, 46, 47, 48, 120
vapor, 65
variables, 7, 11, 12, 44, 47, 51, 55, 57, 58, 59, 61, 67, 68, 71, 72, 75, 76, 100
variations, xii, 19, 55, 56, 59
vector, 61, 93
vehicles, 24
velocity, 36
verbal aggressiveness, 135
versatility, 47
violence, 77, 78, 150, 159
vision(s), 10, 32, 45, 62, 93, 97, 124 151
visual impression, 46
vomiting, 135

W

waking, 13, 152, 153, 155
war, 19, 23
Washington, 59, 161
water, 87, 149, 152
wear, 150
wild animals, 150
wood, 35, 153
wool, 149

Y

Yale University, 166

Z

zoology, 19